THE CLOSE ENCOUNTERS MAN

CLOSE
ENCOUNTERS
MAN

How One Man Made the
World Believe in UFOs

MARK O'CONNELL

DEY ST.
An Imprint of WILLIAM MORROW

THE CLOSE ENCOUNTERS MAN. Copyright © 2017 by Mark O'Connell. All rights reserved. Printed in the United States of America. No part of this book may be used or reproduced in any manner whatsoever without written permission except in the case of brief quotations embodied in critical articles and reviews. For information address HarperCollins Publishers, 195 Broadway, New York, NY 10007.

HarperCollins books may be purchased for educational, business, or sales promotional use. For information please e-mail the Special Markets Department at SPsales@harpercollins.com.

FIRST EDITION

Designed by Michelle Crowe

Library of Congress Cataloging-in-Publication Data has been applied for.

ISBN 978-0-06-248417-8

17 18 19 20 21   RS/LSC   10 9 8 7 6 5 4 3 2 1

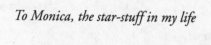

To Monica, the star-stuff in my life

CONTENTS

AUTHOR'S NOTE

This book was born in a basement in Chicago, amid a ring of overfull bookshelves and dented, dilapidated file cabinets. The assemblage of obscure journals, redacted government documents, and plaintive correspondence packing these shelves and drawers was nearly all that was left of the Center for UFO Studies (CUFOS), a UFO research center that had been founded in a Chicago suburb in 1973 by astronomer Dr. J. Allen Hynek.

I had been visiting the archives in search of interesting material for my UFO blog, and on one visit, Dr. Mark Rodeghier, who now stores these antiquated remnants of Hynek's work in his Chicago home, invited me to make use of these ramshackle memory banks to write the definitive account of Dr. Hynek's remarkable dual careers. This was to be the first telling of Hynek's significant accomplishments as an astronomer who pioneered the science of celestial imaging and as a researcher who was on the scene at many of the most amazing UFO encounters in history, and made serious discussions of the UFO phenomenon scientifically—and socially—acceptable.

From a lifetime of studying the UFO phenomenon, I knew of Hynek's pioneering research through his writings, media appearances, and short but notable movie career during his heyday from the 1960s to the 1980s, but when I started to delve deeper into the CUFOS archives, I realized that I understood very little about the man. Hynek had spent nearly twenty years debunking UFO reports for the U.S. Air Force and was considered a sellout, a dupe, and a coward by many Americans for helping the air force "cover up" what they believed the government knew about the UFO mystery. For this sin, he was hated in UFO circles. But he was also ultimately disowned by the air force, because over time his views about UFOs changed dramatically and he began to demand that the government and the scientific establishment conduct a scientific study of the phenomenon, to find out what was really going on. What kind of man, I wondered, could calmly stand at the center of a decades-long conflict and be equally despised by both sides? This is what I wanted to find out.

That drama alone made Hynek a fascinating character to want to write about, but then there was his *other* history—his brilliant yet largely ignored career as an astrophysicist. This was a man who helped win World War II with science, who discovered how and why stars twinkle (not a small thing), and who helped determine the landing sites for the Apollo moon missions. Along the way, he pioneered crowdsourcing, developed the world's first global satellite tracking network, discovered a record number of supernovae, and paved the way for the Hubble Space Telescope. If Hynek's career as a UFO researcher was universally misunderstood and misjudged, his career in astronomy seemed to have been swept under the rug in its entirety. When I discovered that the records of Hynek's scientific triumphs and failures were housed at Northwestern University and the University of Chicago (in significantly better shape than the CUFOS files), and within a few miles of my home, it seemed that the true, complete story of his life and work was demanding to be told. And it came to me as mystically as any of Hynek's career pursuits.

In very little time, I realized that Hynek's two careers informed each other in fascinating ways. Had he not been an astronomer, he would never have found himself at the epicenter of the UFO phenomenon, and had he not been at the center of the UFO furor, he might never have risen to prominence as an astronomer (nowhere was the overlap between Hynek's two careers more dramatic than in his very public, decades-long battle for scientific respectability against astronomer and UFO mocker Dr. Carl Sagan). Furthermore, had his life not followed the exact trajectory that it did, he might never have had an unforgettable cameo in Steven Spielberg's *Close Encounters of the Third Kind,* the research for which was borrowed from Hynek's professional findings. In fact, Hynek and Spielberg corresponded through letters leading up to the production of the film, evidence that not only UFO enthusiasts held Hynek in high regard.

This book, then, is an attempt to rectify Dr. Hynek's story, to find the heroism, humor, and humanity in a man whose name has been relegated to a basement full of file cabinets when it should rightly be written in the stars.

PROLOGUE

We didn't always put dead aliens on ice or pickle them in jars. We used to give them proper burials.

No one knows where he came from—if it was a "he"—and no one knows the exact cause of his death. Nor does anyone know if his soul rests in peace—if he had a soul—but there is a dead alien reportedly buried in the town cemetery in Aurora, Texas.

This one didn't get spirited away to a secret base in Nevada or hidden in a high-security hangar in Ohio. He wasn't subjected to an autopsy and he didn't appear on cable TV. His existence was neither leaked nor covered up. He got a brief mention in a Dallas newspaper item two days after he died, and then the media, such as it was, seemed to forget he ever existed.

Perhaps the spaceman didn't achieve any lasting fame because he met his end in 1897, a full fifty years before the "modern" flying saucer phenomenon began.

At six in the morning on April 17 of that year, early risers in Aurora were witness to an object of pure wonder: a massive, heavier-than-air ship sailing over the center of town, seeming to traverse the sky without the aid of the wind. The wonder lasted only a few minutes, for the metallic airship, apparently suffering from some mechanical malady or the sudden incapacitation of its pilot, began to lose speed and altitude as it careened into the heart of town. As astonished Aurorans scrambled for cover, the airship swooped slowly over the town square and smashed into a windmill on a farm belonging to the town's justice of the peace, Judge Proctor.

The airship was far more fragile than it appeared, however, for the impact with the windmill caused a spectacular explosion and the complete destruction of the strange craft. The windmill was obliterated with it, as was Proctor's flower garden. Miraculously, his cabin was left untouched. Nonetheless, the airship's wreckage was scattered over an area of several acres, according to an eyewitness account in the April 19, 1897, *Dallas Morning News.* "The ship was too badly wrecked to form any conclusion as to its construction or motive power. It was built of an unknown metal, resembling somewhat a mixture of aluminum and silver, and it must have weighed several tons," claimed the reporter S. E. Haydon. "The town is full of people to-day who are viewing the wreck and gathering specimens of the strange metal from the debris."[1] Later, the remaining shards of wreckage were dumped in Judge Proctor's well.

The remains of the pilot were treated with far more respect. If you passed away in the town of Aurora, apparently you were buried in the town cemetery as a matter of course, no matter who you were or where you came from. In that sense the airship pilot was no different from anyone else, and his simple grave was said to be marked by a simple stone—long since gone missing—bearing an etching of his ship. Buried with him were some papers found on his person that bore unknown hieroglyphics—a captain's log of his travels, Haydon supposed. How the citizens of Aurora felt about the airship pilot and his personal effects sharing consecrated ground with their kin was never mentioned in his account, but there must have been some misgivings when people got a look at the deceased. "The pilot of the ship is supposed to have been the only one on board," Haydon wrote, "and while his remains are badly disfigured, enough of the original has been picked up to show that he was not an inhabitant of this world."[2]

The local U.S. Signal Service officer, Mr. T. J. Weems, did not hesitate to offer Haydon his scientific opinion. As an officer of the military branch charged with long-distance coded communication utilizing flags, colored lights, and sometimes even signal rockets, Weems was what passed for a rocket scientist in 1897. Described by Haydon as "an authority on astronomy," among his other considerable skills, Weems "gives it as his opinion that he [the pilot] was a native of the planet Mars."[3]

THE CITIZENS OF AURORA had been hearing quite a bit about airships of late, as the strange vehicles had been sighted all across the country in recent weeks. In fact, four new sightings were reported in the *Dallas Morning News* on the same day the Aurora crash story appeared. The aerial ships took the appearance of flying cigars, elongated ovals, eggs, giant cones, or great balls of light. They were noisy contraptions, powerful enough to fly with considerable speed *into* the wind and to illuminate the ground below them with brilliant electric light. One was even said to have kidnapped a cow. Just as they had caused sensations when viewed by thousands over Sacramento, San Francisco, Omaha, and Chicago, the airships over Texas were the subject of keen public interest. Newspapers had kept the American public informed of the great advances in powered aerial vehicles, but there was not a single working specimen known to exist anywhere on Earth in 1897; what could account for these impossible flying ships?

Perhaps S. E. Haydon saw a way to boost tourism in a region racked in recent years by drought and disease, and contrived to convince all those airship-watching Texans that Aurora was the place to be. "Hayden [*sic*] wrote it as a joke and to bring interest to Aurora," claimed local historian and lifelong Aurora resident Etta Pegues, when interviewed in 1979 for *Time* magazine.[4] Pegues does have a certain credibility on the matter, as she was around, if only four years old, at the time of the incident. Another local authority who was eleven at the time insisted later that T. J. Weems was nothing more than the town blacksmith and that Judge Proctor never had a windmill on his property.[5]

Another, more colorful hoax theory propounded by the *Dallas Morning News* in 1967 suggested that a scrupulously honest employee of the Texas & Pacific railroad company named Joseph

"Truthful" Scully concocted the story of the Aurora airship crash on a lark and found a most efficient way of spreading the tale by word of mouth among the nationwide brotherhood of railroad brakemen. Frank Tolbert, columnist for the *Morning News,* theorized that the hoax succeeded precisely because it had been propagated by a man known for and nicknamed after his integrity (this would not be the last time the name "Scully" would be associated with UFO lore).

Not everyone believed the Aurora crash was a hoax, however. When Tolbert revived the story in the late 1960s, eager amateur investigators located two witnesses who had been alive at the time of the event and recalled hearing vivid descriptions of the airship explosion and the dead spaceman from parents and friends. Two more witnesses were able to identify the exact site of the spaceman's grave, under the crooked limb of an old oak tree on the south side of the cemetery (over the years, multiple requests by investigators to have the alien exhumed have been politely but firmly denied by the cemetery association, by the courts, and sometimes by armed Aurorans).

Besides, one couldn't easily dismiss the Aurora crash story without considering the airship craze in its entirety. Many people had come to believe that the ships were marvelous inventions created in secrecy by enterprising adventurers, much like the brilliant, reclusive airship inventor Robur in Jules Verne's popular 1886 novel, *Robur the Conqueror.* If Robur could build a self-propelled, heavier-than-air flying craft on a secret island, as Verne imagined, perhaps by now, a decade later, someone of exceptional genius had actually succeeded in conquering the air (but was as yet too shy, or too cautious, to reveal his or her invention to the world at large). Many reports seemed to support such a theory: witnesses across the country reported having fascinating encounters with airship inventors and their crews, who would periodically land their ships and get out to stretch their legs, as if they had just pulled over at a roadway rest stop. They described the airship occupants as very human, very dapper gentlemen, who spoke perfect English as they described their travels and hinted at greater wonders to come. Only one witness—the same one, strangely enough, who lost a cow to an airship—was known to have been repelled by the airship occupants, describing them as "hideous people" and "the strangest beings I ever saw."[6]

Which brings us back to T. J. Weems and his conviction that the Aurora spaceman was from Mars. In truth, it didn't matter much whether Weems was a U.S. Signal Service officer or the town blacksmith, for he would likely have come to the same conclusion in either case.

In 1897, it was a popularly held belief among sane, educated, sensible people of Earth that intelligent beings existed on the planet Mars, our closest planetary neighbor. It seemed a perfectly reasonable idea because it originated from perfectly reasonable scientists, such as Stéphane Javelle of the Nice Observatory in France. In 1894, Javelle observed a bright flash on the surface of Mars and speculated that it may have been a signal directed at Earth, or perhaps the firing of a gigantic gun, similarly aimed.

Because Javelle was "already well known for his careful work," the editors of *Nature* declared, "The news therefore must be accepted seriously."[7] *Nature* went on to note, curiously enough, that one natural explanation for the flash would be the planet's aurora.

Mars had first begun to stir the public's curiosity in the summer of 1877, when Milanese astronomer Giovanni Schiaparelli drew attention to a network of dark lines on the surface of the red planet that he described as *canali,* or channels. Schiaparelli never intended to suggest that these lines were anything but naturally occurring features on the planet's surface, but he couldn't forestall the powers of human imagination.

Because the lines on Mars appeared to be so perfectly straight, and in some cases seemed to perfectly parallel one another, American travel writer and hobby astronomer Percival Lowell took it upon himself to translate *canali* as the far more evocative "canals" and to ascribe intentionality to their being. Drawing on his personal fortune, Lowell built an astronomical observatory in Flagstaff, Arizona, for the purpose of conducting the most thorough survey of Mars ever attempted.

Lowell Observatory opened its oddly squared-off dome for the first time in 1894, just as Mars was moving into opposition with Earth. This planetary configuration, which occurs every twenty-six months or so, places Earth directly between Mars and Sol, creating optimum viewing conditions for a Mars-obsessed astronomer. And because his observatory was situated amid the pines and junipers of the Coconino Plateau, near the base of the San Francisco Peaks, at an elevation of 7,250 feet above sea level, Lowell possessed, he claimed, the best view of Mars anywhere on Earth.*

And what Lowell saw on Mars stunned him. Schiaparelli's observations barely began to uncover the wondrous order of the Martian canal system that revealed itself to Lowell. "Their very aspect is such as to defy natural explanation, and to hint that in them we are regarding something other than the outcome of purely natural causes," he proclaimed in his 1895 book, *Mars*. To Lowell, the magnificent canals had clearly been constructed over the course of many years by intelligent creatures and for a very particular purpose: because Lowell and his assistants observed the seasonal waxing and waning of massive areas of what they took to be vegetation, it became clear that the canals had been built as part of a planet-wide irrigation system staggering in its scope. Every Martian spring, Lowell believed, melting ice from the polar ice caps was channeled by the canals to the temperate middle latitudes of the planet, where it brought life to the crops the Martian inhabitants needed to survive.

It was not a sanguine world that Lowell observed. Mars's atmosphere was desperately thin, and with every orbit of the sun

---

* Other observatories had larger telescopes, but only Lowell's was perched where it could get "as good air as practicable, at Flagstaff," according to the preface to his book *Mars*. "A steady atmosphere is essential to the study of planetary detail: size of instrument being a very secondary matter," he went on, apparently in all seriousness. "A large instrument in poor air will not begin to show what a smaller one in good air will. When this is recognized, as it eventually will be, it will become the fashion to put up observatories where they may see rather than be seen."

more of its life-supporting water vapor drifted out of its grip. The canals, then, were a last resort for the Martians, a bold, death-cheating attempt to harness the last remaining moisture on their slowly evaporating world.

"The process that brought it to its present pass must go on to the bitter end, until the last spark of Martian life goes out," Lowell warned. "The drying up of the planet is certain to proceed until its surface can support no life at all. Slowly but surely time will snuff it out. When the last ember is thus extinguished, the planet will roll a dead world through space, its evolutionary career forever ended."[8]

The drama and nobility of it gripped the public. As the turn of the century approached, "it was widely believed . . . that positive evidence had been found for the existence of intelligent life on earth's neighboring planet Mars. This belief stirred an extraordinary controversy that for a time involved much of the literate world," wrote Lowell biographer William Graves Hoyt. "Eminent scientists, philosophers, and moralists joined in the unprecedented debate, while ordinary people everywhere followed its course in books, at lectures, and through lengthy and detailed accounts in newspapers and magazines of the day."[9]

"You grow up with the romance of Mars, you know?" science fiction author Ray Bradbury said when asked years later about man's fascination with Mars in the early twentieth century. "When I was a kid, some of the earliest clear photographs of Mars were being published by the Lowell Observatory in Arizona, and since it's the nearest planet . . . we've always had this feeling that someday, if we went anywhere, it would be to the moon and then to Mars."[10]

Unless, of course, the Martians came here first.

"No one would have believed in the last years of the nineteenth century that this world was being watched keenly and closely by intelligences greater than man's and yet as mortal as his own," H. G. Wells warned readers on the opening page of his science fiction masterpiece *The War of the Worlds*.[11]

"Yet across the gulf of space . . . ," Wells wrote, "intellects vast and cool and unsympathetic, regarded this earth with envious eyes, and slowly and surely drew their plans against us."[12]

To imagine, one short terrestrial year after Percival Lowell made his case for the existence of intelligent life on Mars, that inhuman creatures from the red planet may be intent on our destruction was Wells's masterstroke. Surely a public that wondered so fervently about a civilization on Mars must also harbor fears of what that alien life might be like: whether it looked like us, whether it reasoned the way we did, whether it had the same feelings . . .

Wells did little to allay their fears. His Martians, those "vast and cool" intellects, were little more than heads: slimy, leathery craniums with cold, staring eyes and V-shaped slits for mouths. Supporting the giant heads was a network of slithering tentacles that functioned as arm, leg, and hand all at once. And they brought a most terrible weapon: a heat ray that could incinerate every man, woman, and child on Earth. Eighty years after Mary Shelley's science fiction novel *Frankenstein; or, The New Prometheus* implanted into the cultural subconscious the terrifying thought that a creation of science could turn on its master, Wells suggested that scientific death could rain down on us from the skies.

And the public devoured it. Science fiction author and historian Brian Aldiss credits a good share of the success of *The War of the Worlds* to Wells's deft exploitation of evolution and microorganisms, two of the hot scientific issues of the day, but a lot also had to be said for Wells's "good psychological timing."[13]

"The new journalism was bringing word of the solar system to Wells' public, while Mars in particular was in the general consciousness," Aldiss wrote, noting that Lowell's "speculations about the 'canals' and possibilities of life"[14] on Mars were a boon to Wells's novel. After the triumph of *The War of the Worlds,* in fact, Wells acknowledged that he was "greatly indebted" to "my friend, Mr. Percival Lowell,"[15] whose studies helped Wells to better imagine what type of life might evolve in the peculiar conditions on Mars.

Egged on, Lowell pushed his theories to the limit, swaying the editors of the *New York Times* with his impressive aggregation of observational data. In its December 9, 1906, issue, the *Times* quoted Lowell, now described as a professor and "the greatest authority on the subject" of Mars, as saying that "there can be no doubt that living beings inhabit our neighbor world."[16] Commenting on the obviously manufactured Martian canals, the *Times* wittily quoted the early proponent of intelligent design and clockwork creation, the Reverend William Paley: "A thing made predicates a maker."[17]

The cross-pollination of science and science fiction in the early 1900s, exemplified—and in some ways triggered—by Lowell's and Wells's symbiotic validation of each other's work and speculations, may seem absurd and alarming to modern sensibilities, but it was a rampant force at the time, unencumbered by logic or reason.

Especially where the planet Mars and the solar system were concerned, the barrier between science and science fiction had become infinitely permeable, to the point where fantasists felt free to propose scientific theories, and scientists, who were much less cautious back then, felt free to let their imaginations run wild on the pages of popular newspapers and magazines. In that world, at that moment, science and imagination could safely hold hands in public.

One liberal-minded scientist was unable to resist the temptation to engage in some supremely fantastic thinking in the *Salt Lake Tribune*. "A new and exceedingly interesting theory concerning the life on Mars has been put forward by Professor William Wallace Campbell, of the great Lick Observatory, California," the *Tribune* raved. "He suggests that all life on Mars has taken a vegetable form."[18] Under the bold headline "Mars Peopled by One Vast Thinking Vegetable," the paper devoted nearly a full page, luridly illustrated, to Campbell's fantastic imaginings of an intelligent global fern with an insatiable thirst for knowledge of the universe, and one very unusual appendage.

"The white spot which we sometimes see on Mars is not a pile of snow, but really an 'eye,'" Campbell claimed. "Supported on a tenuous flexible transparent column, it can raise itself miles above the surface of the planet and watch the operations of its vegetable body at any point."[19]

When it wasn't busy monitoring itself, the great Martian vegetable used its enormous telescoping eye as a combination of an astronomical observatory and a gigantic Peeping Tom. "The great 'eye' makes observations of the earth, sun, planets, stars and the whole universe," Campbell declared. "From its vast side it is able to see more and farther than all the telescopes of our earth put together."

"This theory," the *Tribune* decided, "is one of the most plausible that has been put forward."[20]

THE INTOXICATING MIX OF FEAR AND FASCINATION that defined the public's infatuation with science and with Earth's heavenly neighbors reached a crescendo with the approach of Halley's Comet in the spring of 1910. Here was a scientific phenomenon that nearly every human being on Earth could see with his or her own eyes in the night sky—the ultimate shared experience—and that a great many human beings feared. For while the exact time and location of a comet's appearance could, by 1910, be accurately predicted by science, its actual nature and purpose was still clouded in myth and superstition.

It was established fact that Halley's Comet appeared in our skies about every seventy-six years and established lore that its appearance—the appearance of any comet, for that matter—brought about certain calamity and suffering. In 1066, most famously, Halley's Comet foretold the Battle of Hastings and the violent struggle for the British throne that ensued, and after that even Shakespeare submitted that the comet was a bad sign for a sitting monarch. This proved to be the case again in May 1910, when King Edward VII succumbed to failing health and passed away only days before the comet's nearest approach.

Even worse, scientists discovered that all humanity was at risk of following Edward to his doom. Not only was it determined that Earth would be passing directly through the comet's tail for a six-hour period on the night of May 18–19, but astronomers at the Yerkes Observatory in Williams Bay, Wisconsin, utilizing the new technique of spectroscopy to determine the temperature and chemical composition of a luminous body by analyzing the spectrum of light it emits or reflects, found the tail to contain a deadly substance: poisonous cyanogen gas. French astronomer Camille Flammarion was distressed enough by the Yerkes findings to declare that "the cyanogen gas would impregnate the atmosphere and possibly snuff out all life on the planet."[21]

Finding themselves stalled out and stranded on a celestial railroad crossing with an express freight hurtling toward them, the people of Earth gave way to fear and prepared for the end.

"Some people took precautions by sealing the chimneys, windows, and doors of their houses. Others confessed to crimes they had committed because they did not expect to survive the night, and a few panic-stricken people actually committed suicide," reported science writers Gunter Faure and Teresa Mensing. The most gullible bought "comet pills," "comet umbrellas," and gas masks, while the most faithful gathered nervously in houses of worship, prepared to meet their maker. Some, intent on going against the grain, were gripped by an inexplicable end-of-the-world euphoria: "A strangely frivolous mood caused thousands of people to gather in restaurants, coffee houses, parks, and on the rooftops of apartment buildings to await their doom in the company of fellow humans."[22]

One of those rooftops was in Chicago, Illinois, although its viewing party took place almost two weeks ahead of the global deathwatch of the eighteenth, and the guest list was rather small. On the night of May 5, Joseph and Bertha Hynek took their five-day-old son, Josef, to the roof of their West Side home to bask in the light of the comet.

What the mood was on that rooftop, one can scarcely guess, but it must have come as some relief to Joseph and Bertha that they and their newborn son survived the fallout of the comet's tail thirteen nights later. Nonetheless, little Josef, who was to be their only child, may have gotten a sprinkling of comet dust that night, because for the rest of his life, his path would be marked, and sometimes defined, by the appearance and movements of unusual heavenly bodies.

Destined to become a trusted spokesman for the space race, a paradigm-shifting pioneer in astronomical imaging, an authority on the study of UFOs, both lauded and reviled, and an unexpected cultural touchstone in the world of science fiction, Josef Allen Hynek could not help but spend much of his life and career out on a limb, reaching for the lights in the night sky. Born into a world where cunning and intelligent Martians built thousand-mile canals and spied on us through giant eyes, where scientific destruction could rain down on us from the skies without warning, where impossible flying ships could crisscross the skies with impunity, and where a spaceman was laid to rest in a small cemetery in north Texas after crashing his flying ship into a windmill, Hynek would, fittingly, grow up to embody the contradictory nature of scientific inquiry and investigation in the twentieth century, with its simultaneous dependence on and rejection of imagination and wonder.

It wasn't just a boy who was born on May 1, 1910, to Joseph and Bertha Hynek. He was a spaceman.

CHAPTER 1

UNDER THE DOME

JOSEF ALLEN HYNEK WAS BORN into a household three generations deep and redolent with dried tobacco. His parents, Joseph and Bertha, shared the South Avers Avenue house in North Lawndale with Bertha's widowed father, Joseph Waska, and her sisters Anna and Mildred. Both families hailed from Bohemia (now part of the Czech Republic), but while Joseph Hynek was born there and then immigrated to the United States with his parents, Bertha was born in Chicago. The two Josephs, Bertha's husband and father, were in business together as Waska and Hynek, manufacturers and purveyors of cigars. Bertha was a schoolteacher.

Living under the same roof with two other Joes—and being lowest in the pecking order among the three—surely presented young Josef Allen with some identity issues, but in every other respect he seems to have had a perfectly adequate and normal childhood, albeit one obscured by an eternal haze of cigar smoke.

It wasn't until he was seven and contracted scarlet fever that the dust and gas from the tail of Halley's Comet began to exert its influence over the young spaceman's life. Forced to spend several weeks in bed, "Joey" soon suffered from an acute case of boredom and "read practically every book available in the neighborhood."[1]

One of those was *Elements of Astronomy,* a textbook written in 1869 by Selim H. Peabody, Ph.D., LL.D., F.S.Sc. "Out of sheer desperation," Bertha resorted to Dr. Peabody's astronomy textbook—which, it must be said, was "written for pupils in the higher grades of public schools"—and read it aloud to her son.

Despite his youth, and Dr. Peabody's prosaic text, Joey took an urgent interest in the heavens and never looked back. "The system, the law and order of things, grabbed me," he said, claiming that he could never remember a time when he didn't want to be an astronomer.[2]

A gift of a Sears, Roebuck & Co. telescope at age ten seemed to seal the deal.

But there was another suitor for the boy's affections. By age thirteen, Josef had developed a flair for writing and was considering a career as an author. From 1923 to 1927 his short stories regularly graced the pages of *Science and Craft,* the literary journal of the all-male Richard T. Crane High School, and they display a keen talent for elaborate plotting, witty characterizations, and philosophical thought.

Just short of his fourteenth birthday, Josef lost his father. Judging by the Cook County death certificate, Joseph Hynek's March 1924 death at German Evangelical Deaconess Hospital was both sudden and dramatic: the family had known of Joseph's heart condition for less than a month, and only two days passed between the onset of acute symptoms and complete cardiac failure.

It seems only natural, then, that Joey's short stories written in that period would reflect an intense curiosity with fate and the unknown. In "The Mystery of Blue Manor," a haunted house turns out to be the secret hideaway of a renegade German scientist named Professor Stock, who has developed "a marvelous apparatus for the transmission of electrical power." The ghostly apparitions in the house turn out to be manifestations of Stock's "marvelous apparatus," and in Joey's imagination the world of the supernatural is transformed into a scientific phenomenon that can be studied, taken apart and understood.

Likewise, in the adventure yarn "The Foreboding Star," a warning of doom from a fortune-teller manifests itself in an all too rational way and unexpectedly saves the life of a daredevil stunt pilot. A prescient line from this story describes the narrator's state of mind as his daredevil friend reveals the strange curse under which he has lived for so many years: "I felt as a group of scientists would feel on the verge of unraveling a mystery."

Beginning as a contributor to *Science and Craft,* Joey was promoted to the position of associate editor, and although he did not pursue writing as a career, his talents as a wordsmith were to come to his rescue on many occasions when his spoken words failed to adequately convey his expansive thoughts.

A little over five years after his father's death, the other half of Joey Hynek's world crumbled. His mother, Bertha, was diagnosed with breast cancer in February 1929, and in May she died at the same hospital at which her husband had passed away. She was only fifty-eight years old.

By this time, Hynek had graduated from Crane High School and was pursuing a bachelor of science degree at the University of Chicago, only thirteen miles from home but a world away from his family tragedy. Finally settled on a career in science, he made the unlikely decision to pledge Alpha Tau Omega, a predominantly athletic fraternity.

According to Hynek's son Paul, the fourth of five children from his father's second marriage, pledging to ATΩ was a strategic move that could have been made only by a refugee from an all-male high school: "[The athletes] would help him meet girls, and he would help their grades." There was, however, a wrinkle. "There were already two brothers in the fraternity named Joe," Paul said, and Hynek did not want to repeat the experience of being the third and lowest-ranking Joe in the household.

Early on in his university career Hynek—now calling himself J. Allen—showed great promise. A paper from an English class, dated November 12, 1928, and entitled "The Development of the Heliocentric Conception of the Universe," reveals the degree to which his scientific philosophy and mode of expression were developing.

A remarkable passage describing seventeenth-century astronomer Johannes Kepler's discovery of his Laws of Planetary Motion seems to reveal the eighteen-year-old Hynek's dawning awareness of a deep kinship with the scientist. "Perhaps it seems strange that a man should have become so absorbed in one subject that he gave his entire life to the formulation of three laws," Hynek wrote in his paper. "But Kepler was a mystic. He lived in poverty all his life, and he cared for nothing but the search for Truth. He held many mystical ideas about the stars such as that they influence the lives of man, and that each planet has a guiding angel that keeps it from straying off in space, but he never lost his clear reasoning powers in metaphysical speculation."

From his insightful musings on the mystical Kepler, it was not a great leap for Hynek to become interested in the teachings of the Freemasons and the Rosicrucians, with their temptations of expanded consciousness and the search for knowledge. As he delved deeper into esoteric thought, Hynek became enthralled with the concept of "occult science" propagated by philosopher and spiritual teacher Rudolf Steiner, founder of the Waldorf education movement. Steiner believed in the objective reality of the spirit world, to the degree that he felt that the world of spirits could be studied with the same rigor with which we study the physical sciences. He distilled occult science into two principles: first, that there is another, invisible world that is very real and yet hidden from our senses, and second, that if we can develop certain "dormant faculties" we can both perceive and enter that hidden world.

To a curious young man newly alone in the world, Steiner's provocative writings offered a feast of concepts, both challenging and liberating: access to a "supersensible" realm of physical reality; an ability to sense past events; even contact with the afterlife. But Steiner was not a showy mystic, given to séances or guided writing; he was, in fact, surprisingly grounded. "The supersensible world appears to us in such a way that it resembles our perceptions of the sense world," he once said in an attempt to demystify his concepts.[3]

It is interesting, then, that as Hynek's curiosity about a more expansive theory of reality was beginning to blossom, his university career began to take on a bit of a miraculous quality. After graduating with a degree in astronomy from the University of Chicago in 1932, Hynek set his sights on earning a graduate degree via a coveted research post at the U of C's iconic Yerkes Observatory in southern Wisconsin, the same place where the poison gases in the tail of Halley's Comet had been discovered.

The Great Depression had hit the University of Chicago hard, and Edwin Frost, director of the Yerkes Observatory, found himself in the embarrassing position of turning down a great many appeals for assistance: "Unfortunately I cannot give you any encouragement whatever that you could find an astronomical position in America," Frost wrote to an Austrian colleague in early 1932. "We are having all that we can to provide the salaries for our present assistants and there is no opportunity to increase the staff here."

And yet, amid this maelstrom of hardship, Hynek's reputation among the faculty as a promising and dependable student helped him secure a position at Yerkes with astonishing ease. When he wrote to Frost in January 1932 requesting an opportunity to work at Yerkes and to be granted a fellowship to pay his way, Frost's response was surprisingly free of financial anxiety: "I recall meeting you more than once and have heard favorable reprots [sic] of your work from Mr. MacMillan and we should of course be glad to admit you to do work in observational astronomy and astrophysics at the Observatory.

"I am glad that you made applications for the fellowship," Frost went on. "While I do not know how many applications there have been from other institutions, it would seem to me that you have the best chance among graduate students at the University."

Frost then followed up with a note to Hynek's mentor at the Chicago campus, Professor William D. MacMillan, stating, "It seems to me that Hynek should have the first chance for a fellowship of anyone on the grounds." Of course, when it seems to the director that something should occur, it generally occurs. Frost's exact words were that Hynek would be "thoroughly satisfactory." He was in.

It was noted by Yerkes staff that Hynek, in his first weeks at the observatory, was a very industrious worker, to the point that he was quickly driving himself to exhaustion. "The doctor said he was run down due to overwork & improper eating," a concerned staffer wrote to Dr. Otto Struve, who had recently replaced the retired Frost as director. "It seems he ate very infrequently and worked half or more of the night." But overwork and malnutrition were not the only dangers faced by a young grad student like Hynek. As beautiful as it was, the brown brick Romanesque observatory building, with its riotous terra-cotta decorations commemorating the zodiac and ancient astronomical mythology, provided dire housing for underpaid students. "The roof leaked and the room became a cold, dark tomb on winter weekends, when the electric power (except to the telescope) and heat were turned off in the building," reported University of Chicago historian Donald Osterbrock. "Water dripped in during summer thunderstorms, and snow drifted in during the winter."[4]

But one man's "cold, dark tomb" is another man's Fortress of Solitude, and as Hynek worked on his measurements of stellar spectra at the "lonely Yerkes observatory on the tranquil shores of Lake Geneva, Wisconsin," he found solace in his isolation.[5]

"You go to this observatory with just a few other people there, and you feel like you might be a monk, looking at the heavens . . . learning the secrets of the universe," said Hynek's colleague Dr. Mark Rodeghier. "You can see how that would lead to spiritual feelings in the right personality."

Night after night under the ninety-foot main dome, Hynek peered into the firmament, studying the ancient light given off by distant yellow-white dwarf stars and forgetting that anything else ever existed, or ever would. Science and mysticism came together every night in the eyepiece of his telescope . . . time vanished, dimensions contracted.

"The whole thing had a sort of mystical quality," Hynek confessed later in life. "One shouldn't say that in connection with science, I guess . . . but I was so utterly absorbed in the life of the observatory that I had hardly heard of Hitler."[6] He did, however, break away from his absorption long enough to marry Martha Doone Alexander in Fayetteville, Arkansas, on Christmas Eve 1932. Although very little information about this romance exists today, it does prove that Hynek's existence wasn't entirely monastic.

It is not difficult to imagine that during his nights of mystical seclusion at Yerkes, Hynek continued to read Steiner and wondered about how he might access the "supersensible realm."

"Rudolf Steiner had discovered the answer," observed writer Colin Wilson. "His early studies of geometry and science had taught him the 'trick' of withdrawing deep inside himself, until it dawned on him that the inner realm is a world in itself, an 'alternative reality,' so to speak."[7]

Could Hynek's long, patient studies of stellar spectra have taught him to withdraw into a world within himself?

He admitted as much when he recalled sharing his inner contemplations with his boss: "I once asked Struve, riding from Chicago to Yerkes Observatory, when I was a graduate student, 'I wonder what the human race would be like if we had developed emotionally rather than intellectually? If our whole emphasis had been on emotional development, and had developed into feelings rather than trying to probe the chemical nature of the universe.' 'Oh,' he snapped, *'It would be much better if you didn't think of such things!'*"[8]

It is surprising that Struve would become Hynek's mentor and father figure during his time at Yerkes, for it was he who exposed his impressionable young protégé early on to an egregious case of scientific shenanigans. In Hynek's second year at Yerkes, the University of Chicago had taken on a significant role in preparations for the 1933 Chicago World's Fair: A Century of Progress. Slated to open exactly forty years after Chicago's first World's Fair, the 1893 Columbian Exposition, the 1933 event was to be an unbridled monument to speculative imagination, civic pride, and technological prowess, covering nearly 140 acres and costing more than $100 million. Knowing that the Century of Progress was expected to top all other World's Fairs in scope and attendance and that the prestige of the city of Chicago was at stake, Struve's predecessor, Edwin Frost, pitched a grand concept for an opening night extravaganza that could never be equaled.

Frost's idea drew as its inspiration the star Arcturus, a red giant 1.5 times more massive than our sun. Although its light is a dusky orange rose, and it has long since passed its peak luminance, Arcturus shines with an intensity that dominates the heavens; it's the brightest body in the constellation Boötes, the herdsman, and the most brilliant star in the northern celestial hemisphere. It is also quite unruly; instead of rotating with our arm of the Milky Way, like any other good star, Arcturus pursues a contrary course, speeding laterally through the disc of the galaxy as if straining to escape the cosmic order.

Frost's plan centered on a little-known fact and a crude but rapidly evolving technology. The fact was this: Arcturus was estimated to be forty light-years from Earth, so the light it would cast on the Century of Progress had actually been shed by the star during the Columbian Exposition of 1893. The technology was this: a scientific wonder called the "photoelectric cell" that could magically transform light into electricity.

Frost proposed that the staff at Yerkes would place a photoelectric cell at the focal point of its massive forty-inch refractor telescope—the largest of its kind in the world—which occupied the main dome of the observatory. The cell would gather the forty-year-old light from Arcturus*—thus linking this World's Fair to its illustrious predecessor—and convert it to electricity. The electrical energy would be amplified by equipment supplied by Thomas Edison's General Electric Company (with backup facilities supplied by rival Westinghouse Electric and Manufacturing Company), and then sent on a hundred-mile journey from Williams Bay, Wisconsin, to the Chicago lakefront over telegraph lines supplied by the Western Union Company. The electrical impulse, snared by science from a rebellious star and delivered by

---

* Years later it was determined that Arcturus is only thirty-seven light-years from Earth.

an odd partnership between corporate America and the cosmos, would power the switch used to fire up the lights on the first night of the fair.

It was a brilliant conception, but it had one potentially fatal weakness: a single passing cloud, the astronomer's perpetual pest, could blot out the light from Arcturus, scuttling the whole affair and leaving the fair's opening night crowd in bewildering, unscientific darkness. Frost had an answer for this as well: on May 27, 1933, the first night of the fair, telescopes at Harvard University in Cambridge, Massachusetts; Allegheny Observatory at the University of Pittsburgh in Pennsylvania; the University of Illinois in Urbana; and the newly built Adler Planetarium in Chicago would all be directed at Arcturus, ready to send a burst of star energy to the fair via Western Union if Yerkes could not deliver.

The night of the opening, all hands were on deck at Yerkes. Frost had by now retired, and Struve was committed to carrying out Frost's plan to the letter. An immense amount of pride and sacrifice had gone into the Arcturus project, and everyone at the observatory, professors and graduate students alike, had a stake in the success of the endeavor. Science had to triumph, and it had to triumph from the lead observatory and no other; there was no alternative.

One hundred miles away, a crowd of thirty thousand had gathered outside the Hall of Science at the fairgrounds to witness the spectacle of the star beam. The festivities began with "a parade of five hundred persons dressed in national costumes and carrying flags of forty different nations,"[9] followed by a concert given by the Chicago Symphony Orchestra and a 2,500-voice chorus, and a rendering of the national anthem by opera star Lawrence Tibbett. The program was broadcast far and wide by Chicago's clear channel radio station WGN. Rousing speeches were delivered by the exposition board president Rufus Dawes, Adler Planetarium director Philip Fox, and Dr. Frost himself.

Clouds gathered over the fairgrounds and a few drops of rain fell, although the crescent moon was able to break free. A minute later Mars and Jupiter appeared, but where was Arcturus? "The waiting audience was lifted to a tension of awe as the supreme moment drew near," wrote one inspired reporter. "Prof. Philip Fox, master of ceremonies, compared it with the scene when people waited for the Delphic oracle to speak."[10]

IF THE TENSION WAS RISING at the Chicago lakefront, it was becoming unbearable beneath the main dome at Yerkes. Thick clouds were drifting over Williams Bay as well, and Otto Struve was leaving nothing to chance. He made sure one of the graduate students was standing by with a high-powered flashlight; if Arcturus was blocked by clouds at the wrong moment, the flashlight beam would be trained on the inside of the dome and the telescope pivoted to catch its rays instead of those of Arcturus. The photocell in the telescope would never detect any difference . . .

Back at the fair, Philip Fox didn't let the clouds hamper his performance. Standing beneath a towering sign showing the locations of Harvard, Allegheny, U of I, and Yerkes all linked to Chicago, he called out to each observatory in turn to ask if it was ready. Each site radioed back the affirmative, and each signal was met with a rising hum of electricity and a new streak of light across the sign.

At precisely 8:15 Central time, the moment Yerkes reported clear skies, Fox shouted, "Let's go!" and in the next instant science and magic fused into one mystical conflagration. "It was as if a miracle had happened," raved the *Tribune*'s reporter, who declared this the "most daring achievement of science, producing light that never before was seen on sea or land."[11]

"In the center of the flaming circle a star flashed out," reported one fair official. "The switch was thrown, and a searchlight at the top of the Hall of Science shot a great white beam across the sky. It circled slowly from one exposition building to another and, at the touch of its finger, one after another burst into full brilliant illumination."[12] It was a moment of jubilant wonder that left the crowd grasping for new means of expression. "Golden temples came into being, walls of rainbow hue, tubes, arrows and towers of light. It spread over the grounds like a prairie fire," wrote the reporter for the *Chicago Sunday Tribune*. "The voice of the people rose in an ecstatic sound that shut out the hum from the star. Song that had welcomed the light now sped it on its way."[13]

At the observatory, the relieved graduate student hung the flashlight, unused, back on its hook. The scientists at Yerkes had pulled off a miracle of stagecraft and heaved a collective sigh of relief. After forfeiting so much valuable telescope time to the World's Fair, and coming so close to failing at their task, everyone was anxious to get back to more productive work. Struve had the photoelectric cell removed from the telescope that very night and rebuffed requests from the exposition board the following day to repeat the Arcturus stunt every night of the fair.

No one was more relieved to be done with Arcturus and the fair than Hynek, who was recruited by Struve to hold the flashlight aimed at the inner dome of the observatory that night and who came so close to aiding and abetting an act of scientific deception.

This piece of evidence comes from science writer Richard Panek, who in 1977 took an introductory astronomy course taught by Hynek. According to Panek, the astronomer once told a curious anecdote about a telescope and a flashlight: "There was a telescope being brought online and it was a big deal at the time," he said. Hynek told his students that just in case of cloud cover when the big moment arrived that night, *someone*—"I think he said 'I,'" Panek recalled—had been standing by with a flashlight. "I'm almost positive he was referring to himself," Panek said. "It was just this human detail that stayed with me."

Just so the full import of his precarious situation wouldn't be lost on his students, Hynek "gave a little gesture to suggest that [the flashlight] was being kept out of view or being palmed," Panek said.

THERE WAS ANOTHER INSTANCE OF "STAR-SWAPPING," however, that young Hynek not only supported but instigated. For, despite appearances, Hynek was not afraid to ruffle the feathers of the powers that be in pursuit of scientific truth.

This willingness—bordering on eagerness—to go against the grain manifested itself when he was preparing to publish his doctoral dissertation in the University of Chicago's prestigious *Astrophysical Journal.* Barely situated in his first professional post after leaving the secure confines of Yerkes, Hynek used his paper to prove that astronomers at Mount Wilson Observatory in California had misclassified an entire group of bright "F0" stars as "A" stars.

But Hynek had made a political error, and it was a big one: the scientists of Mount Wilson, many of whom had begun their careers at Yerkes, were royalty, the elite wizards of the astronomy world; one did not question their findings lightly. Otto Struve, as editor of the *Astrophysical Journal,* recognized the merit of his erstwhile student's work, but he had serious qualms about Hynek's effusive mode of expression. "The most important suggestion is one of general policy," Struve wrote to Hynek in February 1936. "The article as it now stands reads as a criticism of the work of the Mt. Wilson Observatory, which should by all means be avoided. In the first place, would it be undiplomatic for you to publish anything that could be construed as disrespectful to Dr. Adams; and in the second place I believe that it is not a question of whose classification is right and whose is wrong."

There was a larger issue at stake, as Struve labored to explain: it didn't matter so much whose findings were *right,* but rather whose findings were *more convincing.* He urged Hynek to state, as indirectly as could be contrived, that his findings were more convincing than those of his colleagues and leave it at that.

Hynek accepted with grace Struve's lesson in the qualitative malleability of scientific facts, but he declined to make the edits himself. "Please feel free to make whatever other changes you feel necessary for either the sake of style or policy, especially as concerns Mt. Wilson," he wrote back. "I am afraid I am apt to be too thoughtless of possible consequences in making statements and appreciate your criticisms."

But thoughtlessness was not Hynek's problem. It was the opposite extreme that was to cause him so much trouble.

CHAPTER 2

UNUSUAL STARS

ON A DECEMBER NIGHT IN 1934, an amateur astronomer in England named J. P. M. Prentice detected a suspiciously bright spot in the constellation Hercules. Prentice's spot was to become known as Nova Herculis, the first known stellar explosion to encompass two stars at once. It became one of the most spectacular novae of the century and was said to have inspired Superman's origin story, in which the superlad is sent to Earth just before his home planet Krypton explodes in spectacular fashion. Astronomers the world over rushed to observe and study Nova Herculis as the brilliant yellow pinprick of light quickly grew in magnitude, reaching a point where it was clearly visible in daylight, and it began to exert its pull on Hynek's destiny.

"The appearance of this Nova . . . has made available much more data, both visually and spectroscopically, than any previous appearance of a Nova," reported the *Harvard Crimson* of Prentice's discovery. "This may lead to an explanation of what takes place in the interior of a star. Most stars are not in such a state of flux, and conceal their interiors with clouds of luminous gases, but Nova Herculis represents a remarkable example of stellar activity."[1]

The data referred to in the *Crimson* article—like that captured by the astronomers at Yerkes from Halley's Comet years earlier—was derived from a spectrograph, which analyzes the chemical composition of the light emerging from a luminous body. In short, if your observatory didn't have a spectrograph in 1934, you were missing out on the astronomical bonanza of the decade.

The Perkins Observatory, opened in 1931 in Delaware, Ohio, was one such underequipped facility. Operated by Ohio Wesleyan University in conjunction with The Ohio State University (OSU) in nearby Columbus, Perkins had not yet acquired a spectrograph and had to make inquiries of other observatories to see if such an instrument was available for loan. A spectrograph was located at Yerkes Observatory, and an urgent request was made to borrow the instrument while Nova Herculis was still casting its glow.

Otto Struve struck a bargain with the director of Perkins: he would send the spectroscope, and as a bonus the apparatus would be accompanied by a progression of young graduate students proficient in its operation, Hynek among them. In return, the Yerkes students would receive significant amounts of additional telescope time at Perkins, and Struve would have access to any and all data they collected. It was a phenomenal windfall for Struve, but it was not without consequence, for at the end of the arrangement he lost one of his best students.

"In the freezing cold of a bitter winter, Hynek took spectra of this object night after night, and so impressed the Perkins observatory director that he was recommended for a [teaching] post at the associated Ohio State University," reported the *New Scientist*.[2] Hynek had once mentioned to Struve that he was interested in the teaching opportunity, and Struve had scolded him: "Don't tell that to anybody around here! You'll never get your degree!"[3]

"At that time the thought of any professional astronomer descending so low as to popularize or teach was unheard of," Hynek explained. "Teachers were looked down upon as second-class citizens."[4]

In the end Hynek chose the second-class life of a teacher at Ohio State (with additional duties at Ohio Wesleyan), perhaps because it meant he could continue his work in spectrographic analysis with no interruption. According to science writer Ian Ridpath, "His work revealed that there were many hitherto unsuspected double stars identifiable only from the superposition of their spectral lines—he [called] them 'spectrum binaries.'"[5]

Settling quickly into his new position as associate professor of astronomy, Hynek set about defining himself in a way that suggested he had shed his fascination with the supersensible realm, and rather dramatically at that. In one lecture given at Perkins Observatory in February 1936, Hynek discussed the topic of "Fact and Fancy in Astronomy," cautioning his audience that "science can be considered as a game and a game played by very strict rules."[6]

"The sky and stars have always been symbolical of that which is above human comprehension and this circumstance has led to a great deal of unscientific speculation," Hynek said, identifying the chief challenges to reason: the popularity of astrology and ideas that sunspots, comets, meteors, and eclipses can foretell events or influence human behavior. "When any sense can be made of the contents of such theories they are completely scientifically unsound."[7]

One can sense the influence of Otto Struve in Hynek's abandonment of speculative theories now that he was gainfully employed in academia. After all, it hadn't been that long since Struve had scolded Hynek for his fanciful thinking while the two were driving to Yerkes one night. It was all well and good, Hynek now seemed to realize, to surrender to mystical experience while working in seclusion under the dome of the observatory, but it was quite a different matter when one was a low-ranking academic shaping the minds of young scientists at an institution of higher learning.

ON THE NIGHT OF OCTOBER 30, 1938, the Martian race unleashed its full fury on humanity. The attack was led by Orson Welles, a young radio entertainer who, with his troupe of writers and performers, enacted a fictitious invasion of Earth by the planet Mars that terrified thousands of Americans and gave new life to H. G. Wells's 1898 novel, *The War of the Worlds.*

Amid last-minute fears that the adaptation of the novel was a disaster and that the broadcast was shaping up to be a colossal dud, Welles and his "innocent little group of actors"[8] at the Mercury Theatre on the Air had no choice but to commit to the drama and sell it for all it was worth. "The Mercury did quite consciously attempt to inject realism into *War of the Worlds,* but their efforts produced a very different result from the one they intended," reported Smithsonian.com. "The elements of the show that a fraction of its audience found so convincing crept in almost accidentally, as the Mercury desperately tried to avoid being laughed off the air."[9]

That "fraction" of the audience consisted of thousands of listeners who were convinced—even if only for a few minutes—that they were listening to live news reports of a Martian invasion of Earth taking place in real time. Four decades after H. G. Wells and Percival Lowell had inspired each other to create a gripping vision of doomed, vengeful monsters from Mars, it was a simple enough trick for Orson Welles and the gifted dramatists of the Mercury Theatre to turn that fiction into one of the most famous and controversial media productions of the twentieth century.

Despite screaming headlines that Welles's broadcast had stirred "TERROR THROUGH U.S." and left "RADIO LISTENERS IN PANIC," some historians now believe that the show's impact that night was grossly exaggerated by newspaper editors who were feeling their own terror over the massive loss of advertising revenue to the relatively new medium of radio. "The papers seized the opportunity presented by Welles' program to discredit radio as a source of news," said writers Jefferson Pooley and Michael J. Socolow. "The newspaper industry sensationalized the panic to prove to advertisers, and regulators, that radio management was irresponsible and not to be trusted."[10]

Just the same, an influential 1940 research paper by Princeton psychology professor Hadley Cantril reported that an astonishing 6 million people listened to the Martian broadcast. Of those 6 million, Cantril found that 1.7 million had believed that the invasion was really happening, and 1.2 million of those were "frightened or disturbed" by the show.[11]

Whether the truth of the matter is that the broadcast panicked millions, thousands, or dozens of suggestible listeners, what is known is that it has had an enduring impact and a far-reaching effect on the American psyche. Not only did Welles's *The War of the Worlds* broadcast motivate the Campbell Soup Company to sign on as sponsor of the struggling Mercury Theatre on the Air—saving it from the threat of imminent cancellation and paving the way for the bombastic, self-promoting Welles to transition to Hollywood and set the movie business on its ear with *Citizen Kane*—but the concept of intelligent Martians raining death upon us with heat rays fired from towering tripod war machines was given new currency in popular culture.

WHILE RADIO LISTENERS WERE PANICKING IN THE STREETS—or not—Hynek was moving up in the ranks at OSU to assistant professor of physics and astronomy. His growing popularity on campus is hinted at by his inclusion in a puff piece in the Ohio Wesleyan newspaper about the personalities of the faculty's pets. Noting that Hynek had named his Scottie "Phoebe," after the ninth moon of Saturn, which travels backward, the article quoted Hynek as saying that the pooch "walked backward as a pup."[12]

Career advancement and recognition coincided with a setback, however, as 1939 marked the end of Hynek's marriage. For much of his time at the Perkins Observatory, Martha Hynek worked side by side with Allen as his secretary and handled much of his correspondence. A typed letter to Struve from July of that year, however, bears a handwritten message from Hynek apologizing for the sloppy typing with the telling explanation *"new secretary—just being broken in!"*

Admitting the union was "a bit of a mistake," Hynek and his wife divorced amicably.[13]

Hynek, now less of a recluse and aware of who Hitler was, watched in dismay as America was forced to enter the Great War, and scientific study was hobbled on a near-global scale. Able-bodied scientists were called up by the score for service; Yerkes Observatory lost most of its science staff and was converted into what Struve described as "a plant for war research." Hynek—deferred by the draft board—barely had time for astronomy: "Most of my time now is spent teaching radio and air navigation, or rather, in preparing for the same."

In the midst of the mayhem, Hynek won the love of an undergraduate coed fourteen years his junior. "Mimi Curtis is now wearing a diamond ring, engagement taken from Allen Hynek, assistant professor in the astronomy class," read a February 6, 1942, announcement in the social column of the Ohio Wesleyan student newspaper.[14] The two were married on May 31 and honeymooned— fatefully—in Washington, D.C., where the war effort was appropriating any and every resource within its grasp. A friend in D.C. recruited Hynek to put his scientific knowledge to work for his country, and by June 11 Hynek found himself "called to Washington most unexpectedly."

Because of his expertise in deciphering the language of the electromagnetic spectrum, Hynek was a natural choice to join the team of physicists and engineers at the Johns Hopkins University Applied Physics Laboratory outside Baltimore, Maryland, using radio technology to develop the top secret proximity fuze. As reports editor for the project, Hynek safeguarded the data that was generated by the team; he was given a high security clearance and immersed himself in the technical details of the secret device.

Described by some as one of the "first smart weapons,"[15] the proximity fuze used reflected radio waves to determine how close it was to its target. The fuze could thus be set to detonate a charge or warhead upon reaching a certain proximity to the target, rather than upon direct impact. Gunners, no longer required to hit a bull's-eye every time, scored more kills and lived longer as a result. "It blasted hundreds of German V-1 rockets in mid-flight over the English Channel and the newer V-2s over Allied-occupied Antwerp, Belgium," wrote reporter Douglas Birch.[16] According to Ralph B. Baldwin, a scientist who worked on the project, "It shortened the war drastically. And at the end of the war the general staffs of Japan and Germany didn't know what had hit them."[17]

That was catastrophically true in Japan, as proximity fuzes in the noses of "Little Boy" and "Fat Man" triggered the detonation of those atomic bombs at 1,900 and 1,600 feet, respectively, above the target cities of Hiroshima and Nagasaki, thus ensuring maximum eradication. Hynek was not wrong to describe the fuze as "the Devil's own business."

Initially, though, Hynek was enthusiastic about the war effort: "The work in electronics going on in Washington and elsewhere is truly amazing," he wrote to Struve. "I should like to see Radar developed to the point at which we could get meteoric velocities instantaneously!"

But, as a later missive to Struve reveals, the very concept of military research was becoming deeply troubling to Hynek. "Our laboratory here spends ¾ million dollars a day," he wrote. "How inspiring if even part of that could be directed into constructive channels instead of weapons of destruction. It is my privilege to observe the spending of these sums, and my stomach turns around to see, at times, what waste there is. Often seeming waste is necessary to the interests of time—yet I have seen decisions to spend many thousands of dollars on one small experiment made without thinking through the problem."

"Yet," he added without conviction, "progress is being made."

Yes, progress was being made, and despite his misgivings Hynek understood why: when unlimited talent, time, and resources were focused relentlessly on a problem, there was no end to what human beings could will into being.

"The mechanism used here to gather scientists from many states to concentrate on one problem has proved so productive that the same procedure should be tried in post-war times, aimed at constructive rather than destructive ends," he wrote. "Thus—what progress could be made on cancer, public health, nuclear structure—yes, even corporate objectives—if, for a limited time, the concentrated effort of men loaned from various universities, gathered at one university that provided the nucleus of the project, were applied to it. Certainly in astronomy there must be far more planning on joint problems."

It was with tremendous relief that Hynek resumed his teaching duties at Ohio State and Ohio Wesleyan after his government service came to a close with the end of the war. Now in his late thirties, he had acquired a small measure of scientific stature; in addition to receiving a citation from the Bureau of Ordnance of the Navy Department for his work on the proximity fuze, he was named associate professor and director of OSU's antiquated Mc-Millin Observatory (in addition to his duties as director of the Perkins Observatory). Moreover, he had by now accumulated a respectable number of scientific publications, most dealing with the measurement of stellar magnitudes through the use of stellar spectra and the study of binary stars and novae.

His family had doubled in size during the war years—Scott had been born in October 1943, and Roxane followed in May 1945—and domestic life seemed much better the second time around. According to son Paul, Mimi was a perfect match for his dad: "She was very grounded, also from the Midwest—northern Ohio—but also very intelligent and like my dad very intellectually curious.

"My mom was a college dropout who married a divorcé college professor—very risqué," Paul went on. "She was a home economics major, she had never lived by herself, and here she is, thrust into this academic milieu from the other side of the fence, and she was overwhelmed at first. But she quickly got her feet and her strong will prevailed."

The official bio prepared by the Ohio State University News Bureau reveals the highlights of Hynek's world circa 1946–1947: he enjoyed tennis, swimming, and table tennis and was an avid ham radio operator; and he had served as a visiting professor at Harvard in the summer of 1941 (this was no small accomplishment for a junior faculty member from a small college—he happily wrote to Struve that his stint at Harvard should provide "a most enjoyable and profitable time").

In his new post as associate professor, Hynek was tasked with building a stronger astronomy program within the OSU Physics Department, one that would rival programs in Michigan, Wisconsin, and Indiana. Invigorated by his return to academia, he tackled the chore with relish. The Perkins Observatory was a wonderful facility, but it had, in Hynek's view, "suffered in the past from lack of a strong department on campus to carry the main brunt of astronomical instruction."[18] Hoping to remedy that situation, he set about broadening the course selections and making basic astronomy classes available to more students.

Hynek believed that the university's educational endeavors should reach out to the community as well, and he quickly developed a passion for sharing scientific certainties with the public. Already a member of the American Astronomical Society, he helped found the Columbus Astronomical Society and hosted meetings at OSU. He began to give talks to science clubs and civic organizations in the Columbus area on a regular basis. He developed an ongoing radio program called *Sky Unlimited* that aired on the campus radio station WOSU, and he published a booklet called *Stars over Ohio* to commemorate the fiftieth anniversary of Ohio State's McMillin Observatory, for distribution to secondary schools throughout the state.

IN THE AFTERMATH OF WORLD WAR II, U.S. Army Ordnance had transported nearly 120 "Peenemünders"—German rocket scientists nicknamed after the secret Nazi base on the Baltic Sea, where early rocketry development had been supervised by Dr. Wernher von Braun—to White Sands Proving Ground in New Mexico. "One of their first tasks was to assist in scientific and military V-2 launches," wrote historian Michael J. Neufeld, adding that von Braun's group played "a historic role in the rise of the guided missile and the space launch vehicle."[19]

The rocket testing program was overseen by the U.S. Army's Upper Atmosphere Research Panel, sometimes referred to as the V-2 Panel, and its goal was to find applications for the German rockets captured at the end of the war to study solar radiation and X-ray astronomy in the upper atmosphere. Hynek was a natural to consult on the selection and acquisition of instruments to be launched in the rockets, and he had a crucial connection: one of the V-2 Panel members was Dr. James Van Allen of Johns Hopkins, with whom Hynek had worked on the development of the proximity fuze during the war.

Van Allen had, essentially, reunited the team that had developed the proximity fuze, and Hynek found the consulting job at White Sands a comfortable fit. He was accorded significant influence in determining the focus of the rocketry program, and he did not waste the moment. While serving the needs of the V-2 Panel, he was also able to pursue his own areas of special interest: using aerial photography for "meteorology and long-range reconnaissance," and observations of "the ultraviolet spectrum of the sun to study the ultraviolet energy distribution in the solar chromosphere and corona."[20]

Three years earlier, Hynek had shared a vision with Struve in which groups of scientists could come together to work on great problems in unison, and here, on the frontier of space exploration, he saw an opportunity to make the dream real. Johns Hopkins had ample funding but was short on staff due to a massive cosmic ray study, while Perkins and Yerkes had plentiful staff but slim resources; why not combine strengths for the greater good? "Dr. Van Allen has accepted my suggestions and has asked me to inquire whether a Yerkes-Perkins group would be interested in sending up a spectrograph on their own, in cooperation of course, with Johns Hopkins," he wrote. "Funds can be obtained from the laboratory—what is needed is a group to supervise the construction of such a spectrograph."

Hynek succeeded in creating a peacetime scientific endeavor that took on the focus and urgency of a wartime effort. As he did so, he was setting the stage for scientific revolution: while demonstrating the scientific value of rocket flight during peacetime, he was formulating the concept of capturing images of astronomical bodies outside the obscuring influence of Earth's atmosphere.

He was also, unwittingly, creating a whole new breed of scientist. According to David DeVorkin, historian at the Smithsonian National Air and Space Museum, "Astronomers who were closely tied to mainstream astronomy were reluctant to commit themselves to the long-term effort of developing astronomical instrumentation for sounding rockets, but they were also uncomfortable with the fundamental nature of the activity."[21]

According to DeVorkin, these "mainstream" astronomers traditionally put great stock in using dependable scientific instruments that had proven their reliability and durability over the years. Not so with Hynek and his pioneering rocketry colleagues, who knew that their instruments would lead very short but productive lives. "Some of the experimental physicists . . . delighted in the prospect of constantly building and improving on their instrumentation, being driven to do so by the fact that their instruments were likely to be destroyed upon use."[22]

WHEN HE WASN'T DELVING INTO UNCHARTED, highly classified scientific waters, or developing astronomical instruments intended to crashland with our first rockets, Hynek was gaining a reputation in the astronomy field for what Struve described as "your studies of unusual stars." Wherever there was a star exhibiting unexpected properties or inexplicable behavior—the confounding velocities of mass escaping Beta Lyrae, the "paradox" of Zeta Tauri's variable hydrogen curve, Phi Persei with its "remarkable" helium lines, the "attractive astrophysical problems" of P Cygni, and the "striking variations in the spectrum" of Gamma Cassiopeiae were especially troublesome and tantalizing—Hynek was either at the telescope making observations or being called in to interpret puzzling photographic plates or spectroscopic readings.

But another, more unusual visitor in the sky was about to make itself known. When it arrived, the apparition was of an altogether different nature from that of Halley's Comet or Nova Herculis. Although, like the nova, its appearance came as a surprise, and it was first spotted by an amateur, its brilliance was fleeting and its true nature is still under debate.

On the afternoon of June 24, 1947, a Boise, Idaho, businessman named Kenneth Arnold was flying his small plane across the Cascade Mountains to Yakima, Washington, when he decided to fly closer to the peaks and keep watch for signs of a U.S. Marine Corps plane that had gone down in that area. As he pulled up out of a canyon, something strange caught his eye—nine somethings, in fact. "I noticed, to the left of me, a chain, which looked to me like the tail of a Chinese kite, kind of weaving and going at a terrific speed . . ." Arnold later reported. "They seemed to flip and flash in the sun, just like a mirror."[23]

Arnold reckoned that the objects were approximately twenty to twenty-five miles away and that they formed a "chain" five miles long. "As I kept looking at them, I kept looking for their tails; they didn't have any tails!" he said, later adding, "I could see them only plainly when they seemed to tip their wing, or whatever it was, and the sun flashed on them. They looked something like a pie plate that was cut in half with a sort of a convex triangle in the rear."[24]

Watching the objects as they wove between the mountain peaks and dipped into canyons, Arnold was able to clock their time between Mount Rainier and Mount Adams. Later, looking over his maps at the airport, Arnold calculated the objects' speed between the two peaks as "around 1,200 miles an hour."[25]

Arnold mentioned the bewildering aerial encounter when he stopped at Yakima, but airport personnel and fellow pilots were dismissive of his claims. It came as a surprise to him, then, that when he landed in Pendleton, Oregon, some time later, his story had been called ahead and a crowd of reporters awaited his arrival. Although the objects Arnold saw were roughly crescent-shaped, his description of their motion as resembling saucers skipping across water was the image that took hold. One reporter dubbed the objects "flying saucers" and, thus, the concept as we know it entered into the public consciousness.

"The phrase allowed people to place seemingly inexplicable observations in a new category," wrote Dr. David Jacobs in *The UFO Controversy in America*. "Witnesses scanning the sky could now report that they saw something identifiable: a flying saucer. Moreover, the term subtly connoted an artificially constructed piece of hardware; a saucer is not a natural object."[26]

Jacobs also noted that Arnold's report encouraged scores of people to come forward with reports of unusual objects seen in the sky. "Many of these sightings occurred *before* Arnold's. In this sense the Arnold sighting acted as a dam-breaker and a torrent of reports poured out."[27]

With his cool head, his eye for detail, his keen powers of description, and his pilot's sense of accurate distance, direction, and velocity, Arnold unwittingly set the standard for the perfect flying saucer witness. He also unintentionally gave rise to the concept of ridiculing flying saucer witnesses, finding that many news reporters treated his story "with amusement and disbelief."[28]

"Resentful of what he termed 'press ridicule,' Arnold retorted 'They can call me Einstein, Flash Gordon, or just a screwball, but I am absolutely certain of what I saw.' He added that if he ever again saw a phenomena in the sky . . . 'even if it were a 10 story building flying through the air' . . . he would not say a word about it."[29]

The press's attitude toward Arnold's unprecedented claims was perhaps understandable, but as more sightings were reported in the summer of 1947, Arnold's story seemed less outlandish every day.

On June 28, a brilliant light zigzagged across the sky over an airfield in Montgomery, Alabama. On July 3, a group of wingless objects cavorted around the sky above South Brooksville, Maine; the appearance of the objects was marked by a loud boom, and they behaved like "a swarm of bees." On July 6, a witness in Warren, Ohio, spotted a bright object in the sky that was going 500 to 700 miles an hour and appeared to land. On July 7 and 8, several witnesses at Muroc Air Force Base in Muroc, California, saw numerous bright spheres and saucer-shaped objects moving about in the sky. On August 19, policemen in Twin Falls, Idaho, saw glowing objects travel at a terrific speed across the sky in a triangular pattern.

All summer long reports kept coming from every corner of the country . . . The trouble was there was not a soul on earth who knew what to do with them.

CHAPTER 3

THE CROWDED SKY

"IT APPEARS METALLIC, OF TREMENDOUS SIZE . . . I'm trying to close in for a better look."[1]

This was the last transmission the control tower operators at Godman Army Airfield received from Air National Guard pilot Captain Thomas Mantell on the afternoon of January 7, 1948.

Mantell, flying at about fifteen thousand feet, had just broken away from two other P-51 fighters pursuing a brilliant aerial object that had been sighted by nervous officers at Godman Airfield, Fort Knox, Kentucky. In the last moments of the chase, Mantell alone was able to maintain visual contact with the massive object, and he was determined not to lose it.

Mantell advised the tower at 3:15 that he was going to approach the target. The radio silence that followed lasted three minutes. In that time, Mantell's fighter could have covered about eighteen miles, but no one knows how close it ever came to the object of its pursuit. The next anyone saw of Mantell's plane, it was screaming down out of the sky in a tight spiral, and in seconds it had disintegrated into a farm field in northern Kentucky.

Mantell's wristwatch was shattered at 3:18.

The air force called it Incident #33. It was the first time on record that a military pilot had lost his life while chasing a flying saucer, and it took the phenomenon, and the air force, into entirely unfamiliar territory.

According to the air force's final report, the affair started at approximately 1320, when Kentucky State Police and Fort Knox Military Police radioed Godman Tower to report "a large circular object from 250 to 300 ft in diameter" had been sighted by multiple witnesses over Mannsville, Kentucky. The observer in the tower, Technical Sergeant Quinton A. Blackwell, verified with Army Flight Service that the object had also been sighted over Irvington and Owensboro, Kentucky, and by 1350 it was visible to Blackwell and the rest of the tower staff.

The object was still in sight from Godman Tower more than thirty minutes later, when a quartet of Air National Guard P-51 Mustangs* approached from the south, en route from Marietta, Georgia, to Standiford Field in Kentucky. After verifying that the army did not have any experimental aircraft in the sky that day, Technical Sergeant Blackwell radioed Flight Leader NG 869, Captain Mantell, and asked him and his group to try to get close enough to the object to identify it.

One of the four planes was low on fuel and broke off, while Mantell and the other two pilots headed south on an intercept tangent. One of Mantell's wingmen famously radioed Godman Tower to ask, "What the hell are we looking for?"[2]

No one was able to answer that question with any degree of certainty. The military police at Fort Knox first described it as "a small white object in the southwest sky. It appeared stationary. Could not determine if object radiated or reflected light. Through

---

* Some reports of the Mantell incident refer to the planes as F-51s, others as P-51s. These are in fact different designations for the same aircraft. The change in nomenclature came about when the Army Air Force was spun off from the army in 1948 and became the U.S. Air Force. Under the Army Air Force, Mustangs were designated "Pursuit" and thus were P-51s. Under the independent air force, the "P" designation was changed to "F" for "Fighter," hence the change to F-51. For clarity, I refer to the planes here using the long-established and more familiar "P" appellation.

binocs it appeared partially as parachute with bright sun reflecting from top of the silk, however, there seemed to be some red light around the lower part of it."[3]

Of the tower crew at Godman Airfield, one witness thought it "resembled an ice cream cone topped with red,"[4] while another said "it would seem that it was at least several hundred feet in diameter."[5]

The commanding officer, Colonel Guy Hix, reported, "It was very white and looked like an umbrella. I thought it was a celestial body but I can't account for the fact it didn't move. I just don't know what it was."[6]

That Mantell was in the area at all that day was a bit of a fluke. He was in command of three other Air National Guard pilots who had volunteered to be flown from Kentucky down to Georgia to pick up four P-51s that had been left behind on a previous exercise. Their mission was simple: ferry the planes back at low altitude to their home base at Standiford Field in Louisville.

When Mantell received the request from Godman Tower to change course to identify the object, he and his flight were nearly home. But perhaps the break in routine was welcome to Captain Mantell, for he wasted no time in turning his fighter around and heading off in pursuit.

There was just one problem: the ferry mission was planned for low-level flight, so Mantell's plane had no oxygen. When he set off after the object, he knew that he was climbing much higher than he should without oxygen, but, as a veteran combat flyer, he had the reflexes and nerves of a fighter. "[Mantell] loved the P-51, felt he was the master of it, and flew . . . not carelessly but like an aggressive fighter pilot," said his best friend, Captain Richard Tyler, in his accident report for the Kentucky Air National Guard. "I firmly believe that if he thought he had any chance of catching this object he would have pursued it knowingly to his death."[7]

The Army Air Force concluded that Captain Mantell passed out due to lack of oxygen at 25,000 to 30,000 feet, at which point his plane leveled out and started its descent. "It then began a gradual turn to the left because of torque, slowly increasing degree of bank as nose depressed, finally began a spiralling dive which resulted in excessive speeds causing gradual disintegration of aircraft which probably began between 10000 and 20000 feet."

In an affidavit filed with the Army Air Force, a William Mayes of Route #3, Lake Spring Road, in Franklin, Kentucky, reported hearing an airplane above making "a funny noise as if it were diving down and pulling up, but it wasn't, it was just circling. After about three circles the airplane started into a power dive slowly rotating.

"It started to make a terrific noise, ever increasing, as it descended," Mayes reported. "It exploded halfway between where it started to dive and the ground."

The wreckage scattered in a north to south line for close to a mile. Crash investigators found that the canopy remained locked, indicating that Mantell made no attempt to abandon his plane, either because he was unconscious during his descent or had suffocated at thirty thousand feet.

He left behind a wife and two young boys, one six years old, the other eighteen months.

"CONFUSION" WAS THE WATCHWORD for the flying saucer phenomenon in those early days. Since Kenneth Arnold's sighting had caught the nation's attention the previous June, all flying saucer reports were submitted to the bewildered and unprepared staff at the Army Air Force Air Technical Intelligence Center (ATIC), under the Air Materiel Command (AMC) at Army Air Force Technical Base in Dayton, Ohio.

But ATIC had never developed any clear guidelines for assessing or responding to the reports, despite the fact that the calls kept coming in. By September 1947, the head of AMC, Lieutenant General Nathan F. Twining, was concerned enough about the massive influx of reports that he sent an unprecedented recommendation to his superiors that flying saucers were "something real and not visionary" and that a detailed, classified study of the phenomenon be authorized.

Twining's recommendation was approved on December 30, and a flying saucer investigation unit dubbed Project Sign was authorized to be set up under AMC. The new project was opened on January 23, 1948, at the newly named Wright-Patterson Air Force Base. It was commanded by Captain Robert R. Sneider, and before the paint was dry on the office-door stenciling, the staff was scrambling to explain what Captain Mantell had been chasing. What had happened to Mantell between 3:15 and 3:18? Had he caught up with the object in that time, perhaps even tried to engage it? Because of those three minutes, flying saucers had suddenly taken on a menacing, deadly aspect, and a calm, decisive response from the government was in order.

Scrambling for calm, decisive responses was to become the air force's stock-in-trade where flying saucer incidents were concerned, and sometimes it worked. In Mantell's case, there was a recent precedent that gave Project Sign what seemed like a convenient opportunity to write the incident off as pilot error. Because an air force pilot had recently chased Venus in a well-publicized incident, it was easy enough to apply the same cause to the Mantell case: he had died while mistakenly pursuing Venus.

The press and the American public found this explanation lacking. To make the Venus story stick, Project Sign needed a professional astronomer to validate its conclusion. But the task of recruiting such an expert was daunting: Where in Central Ohio could the air force find a professional astronomer who already held a high security clearance and could go right to work with a minimum of red tape?

"IN 1948, WHEN I FIRST HEARD OF THE [FLYING SAUCERS], I thought they were sheer nonsense, as any scientist would have," Hynek was to write years later. "Most of the early reports were quite vague: 'I went into the bathroom for a drink of water and looked out of the window and saw a bright light in the sky. It was moving up and down and sideways. When I looked again, it was gone.'"[8]

To Hynek's sensibilities, flying saucers were a distraction from real science and an obvious mass delusion shared by a public that was jittery about another Pearl Harbor: "I had joined my scientific colleagues in many a hearty guffaw at the 'psychological postwar craze' for flying saucers that seemed to be sweeping the country and at the naiveté and gullibility of our fellow human beings who were being taken in by such obvious 'nonsense.'"[9]

Although hearty guffaws were not an official prerequisite for the position of consulting astronomer, Hynek's healthy skepticism only enhanced the many attributes that made him an ideal recruit for Project Sign. And in the spring of 1948, the air force sought him out.

"At the time, I was director of the observatory at Ohio State University in Columbus," Hynek wrote. "One day I had a visit from several men from the technical center at Wright-Patterson Air Force Base,* which was only 60 miles away in Dayton. With some obvious embarrassment, the men eventually brought up the subject of 'flying saucers' and asked me if I would care to serve as consultant to the Air Force on the matter."[10]

Hynek accepted, thinking that it would be quick work to explain away most, if not all, of the sightings as misidentifications of quite ordinary celestial bodies. There were professional risks associated with becoming involved in such a questionable line of inquiry, but because he had accepted an invitation to take part rather than volunteering, Hynek had a built-in safety factor. Besides, it made sense in purely practical terms: "I was a natural choice," he wrote, describing himself as "the closest professional astronomer at hand."[11]

"It was thus almost in a sense of sport that I accepted the invitation to have a look at the flying saucer reports . . ." he went on. "I also had a feeling that I might be doing a service by helping to clear away 'nonscience.' After all, wasn't this a golden opportunity to demonstrate to the public how the scientific method works, how the application of the impersonal and unbiased logic of the scientific method (I conveniently forgot my own bias for the mo-

---

* Wright Field was merged with nearby Patterson Field in 1948, creating Wright-Patterson Air Force Base. This consolidation took place approximately a year after the Army Air Force was spun off from the U.S. Army to create the independent U.S. Air Force.

ment) could be used to show that flying saucers were figments of the imagination?"[12]

To Hynek, then, it was mere coincidence—and a rather happy one at that—that the air force approached him to give Project Sign an aura of scientific respectability. "I was somewhat like the proverbial 'innocent bystander who got shot,'" he insisted.[13] And so a perfect match was born. As described in an August 31, 1949, memorandum for record from the air force, Hynek's role in Project Sign was simple and straightforward:

> Dr. J. Allen Hynek, astronomer, Ohio State University, was awarded a contract, effective 16 December 1948 to 30 April 1949, to determine which reported objects might be attributed to natural celestial phenomenon. Dr. Hynek analyzed the first 244 incidents. Dr. Hynek accepted each case at face value, without discounting evidence that sometimes "verged on the ludicrous" and without taking into consideration psychological factors.

"Hynek was a junior faculty member and happy with the opportunity to make some extra money from the Air Force," observed UFO chroniclers Michael Swords and Robert Powell, adding, "He knew that scientific consultants needed to 'please the boss' to keep their jobs."[14]

Once he had signed on as an official consultant to Project Sign, Hynek started right in pleasing his new boss, Captain Sneider: he examined the evidence in the Mantell case and, without hesitation, agreed with the standing verdict.

"It appears to the present investigator, in summing up the evidence presented, that we are forced to the conclusion that the object observed in the early evening hours of January 7, 1948, at these widely separated localities, was the planet Venus," Hynek wrote in his April 30 report to Project Sign.

"On the whole, Sign's UFO investigations were fairly good," explained Dr. David Jacobs. "Its main problem was that the staff was too inexperienced to discriminate between which sightings to investigate thoroughly."[15]

"In the beginning, UFO reports were vague and sketchy, as I was to learn when I took on the responsibility of trying to explain as many as I could astronomically," Hynek recalled. "ATIC just couldn't get the kind of 'hard data' the military was used to getting; they wanted close-up photos, pieces of hardware, detailed descriptions, and so forth. Instead, a military pilot would report that he saw a metallic-looking object, possibly 'disc-shaped'; a wingless craft which 'buzzed' him and then shot away at incredible speed—and that was about all."[16]

Hynek later recounted several flying saucer reports that he was able to explain away for Project Sign as perfectly natural astronomical phenomena:

"On July 11, 1947, in Codroy, Newfoundland, two people noticed a disc-shaped object moving at a very high velocity and having the size of a dinner plate. The object was very bright and had an afterglow behind it that made it look like a cone." Hynek's explanation: "It is extremely unlikely that the object was anything more than a fireball."

"On March 7, 1948 USAF officers in Smyrna, Tennessee, watched an oval object in a direction WNW from Smyrna. It was yellow-orange in color and moved very slowly until about five degrees above the horizon." Hynek's explanation: "The sighted object here was undoubtedly the planet Venus."

"On November 8, 1948 a weather observer in Panama observed a spherical object with a tail like a comet for forty minutes." Hynek's explanation: "It seems entirely probable that the object sighted was the comet 1948L, which had been discovered two days earlier in Australia."

Hynek even had an opportunity to chime in on the sighting that started it all: "In his review of the Arnold Incident, however," read an air force press release, "Dr. Hynek has come up with what he terms 'certain inconsistencies' in Arnold's estimates of size, speed and performance of his 'saucers.'"[17]

Although he allowed in his report that he couldn't explain the incident away as "sheer nonsense," Hynek contended that the reflections of sunlight on the strange craft that caught Arnold's eye were troubling. For Arnold to have noticed the objects as the result of a series of direct reflections, Hynek reasoned, the objects had to be either far closer than twenty to twenty-five miles, or they had to be extraordinarily huge, perhaps one hundred feet in height. It followed that if Arnold's estimates of the objects' distance and size were in question, then so was his staggering estimate of their speed.

Hynek reported, "It appears probable that whatever objects were observed were travelling at sub-sonic speeds and may therefore have been some sort of known aircraft."[18]

The boss was pleased. Very pleased.

At first glance, Hynek seems to have been rejecting his earlier, mystical aspirations to reach beyond the known, but in fact, he was merely attempting to approach the problem from a new angle. He still felt that every mystery could be explained, but where Rudolf Steiner believed that one had to enter the spirit world to find the explanation, Hynek was now trying to achieve the same goal by drawing the mysterious out into the world of the physical.

"As an astronomer and a physicist, I simply felt a priori that everything had to have a natural explanation in this world," he said. "There were no ifs, ands or buts about it. The ones I couldn't solve, I thought if we just tried harder, had a really proper investigation, that we probably would find an answer for."[19]

As Project Sign continued to receive puzzling reports and to struggle with the paucity of hard evidence, two camps started to emerge. Some staffers—stunned by the discs' apparent wingless flight and unnatural maneuvering capabilities—felt that flying saucers represented an extraterrestrial phenomenon. Others felt that the phenomenon could be explained away in its entirety as misidentifications, hallucinations, and hoaxes. The extraterrestrial theory was in ascendancy. Hynek, however, was on the other side of the fissure.

Of the 243 domestic and 30 foreign flying saucer incidents

studied by Project Sign in the one year that it was operational, Hynek had found that roughly one-third could be explained as misidentifications of astronomical phenomenon, about one-third were misidentifications of man-made objects, a small fraction had to be set aside as having "insufficient information," and 48 cases—20 percent—were left "unexplained."[20]

"My batting average was about 80 percent," he said, "and I figured that anytime you were hitting that high, you were doing pretty good."[21]

What's left out of that statement is that some of the forty-eight unsolved cases were thoroughly baffling, as Hynek's notes reveal. In Incident #71, for example, a sighting in Las Vegas, Nevada, that took place on either October 8 or 9, 1947, a retired air force pilot out for a drive watched what he thought was a skywriter, then realized there was no airplane at the head of the trail of white smoke; in fact, there didn't seem to be anything there at all. The object, whatever it was, was too small to see but moving approximately 800 miles per hour, and it performed a 180-degree-plus turn before disappearing behind a distant mountain.

"In everything but the course flown, the description given here answers to that of a fireball," Hynek's analysis began. "The course indicated in this incident, however, appears almost fatal to such a hypothesis. No fireball on record, to this investigator's knowledge, has been known to turn back on itself . . . To execute a curved trajectory would require highly extraordinary circumstances indeed, and a meteoric explanation for this incident must be regarded as most improbable."

Incident #40 was even more confounding. In this sighting, which took place in Phoenix, Arizona, on July 7, 1947, a private citizen watched as a twenty- to thirty-foot elliptical gray object with a distinct "cockpit" descended at 400 miles per hour, spiraled twice, and then quickly ascended and disappeared. In an apparent first, the witness had a camera close at hand, and the two resulting photos proved to be quite problematic to Hynek.

After declaring "no astronomical explanation seems possible for the unusual object cited in this incident," Hynek made a rather bold and prescient statement: "The present investigator would like to suggest that this incident, #40, being one of the most crucial in the history of these objects, be reopened for investigation. The actual camera used by [the witness] should be examined, and the original negatives preserved . . .

"It is unfortunate that a competent investigator was not dispatched at once to 'reenact the crime' with [the witness] and to obtain sketches of the trajectory, etc., before details faded from his memory . . . Physical data like these are absolutely essential if we are to get anywhere in any basic physical explanation of these incidents."*

Upon considering Incident #122, an April 5, 1948, sighting at Holloman Air Force Base, in which three civilian scientists—trained observers searching the sky for an experimental balloon—witnessed an indistinct circular object that carried out "violent maneuvers" at a high rate of speed and then disappeared before their eyes, Hynek all but gave up, writing simply: "At the moment there appears to be no logical explanation for this incident."

And that's where his official involvement ended. Hynek submitted his astronomical assessments of the cases, as he was hired to do, and then he returned to Ohio State. He did not have any say in Project Sign's policies, research methods, or final report.

His air force colleagues could not afford to be as sanguine as Hynek was about those forty-eight unsolved cases, however. "The Project Sign staff was left with some unsettling implications," Jacobs wrote. "If the objects were real but not Soviet or American, and if their flight characteristics did not match the state of technology at the time, then perhaps they were not ordinary: perhaps they came from another planet. One group at Project Sign began to explore this possibility seriously."[22]

---

* Although Hynek expressed regret over the lack of information about the camera and film used, elsewhere in the blue book report it is noted that the witness lent his Brownie camera, prints, and negatives to air force intelligence for analysis and then had some difficulty getting them back . . . an unfortunate but very real precedent.

When one considers that Project Sign's February 1949 final report included more than a dozen pages earnestly discussing the flight characteristics of flying discs, torpedoes, and orbs, and a further nine pages discussing "the likelihood of a visit from other worlds," with special attention paid to Mars and Venus, it becomes apparent that the extraterrestrial hypothesis caused a significant amount of concern in some quarters.

CHAPTER 4

DEBUNKED

THROUGHOUT THE LAST HALF OF 1947 and the first half of 1948, strange flying objects of various shapes, sizes, and colors had been reported from one end of the United States to the other: there were crescents, discs, spheres, cigars, torpedoes, saucers; there were lights of blue, red, green, yellow, silver, white; there were sizes from a few yards from end to end to hundreds of feet across. But until July 24, 1948, at 2:45 A.M., there hadn't been a flying saucer with glowing windows.

The object sighted that night by the pilots of an Eastern Air Lines DC-3 flying from Houston to Atlanta was unique in several ways. For one thing, it seemed to actively avoid the DC-3 in a way that hadn't been encountered before. For another, it visibly displayed what the pilots took to be its mode of propulsion (although they hadn't a clue what that mode of propulsion might be).

At the time of the encounter, pilot Clarence Chiles and his copilot, John Whitted, were flying twenty miles southwest of Montgomery, Alabama, at five thousand feet, under a clear sky and a full moon. According to the Project Sign case report on the Chiles-Whitted Incident, the encounter began when the pilot and copilot sighted "a dull red exhaust some 700 feet ahead, a little above and to the right of the airliner."

The report states that Chiles turned to his copilot and re-marked, "Look, here comes a new Army jet job." A few seconds later, Chiles knew he must be wrong. The object turned slightly to its left—indicating that it was under intelligent control—approached the right side of the DC-3, and streaked past an instant later, at a distance of no more than half a mile. Whitted, in the copilot seat, saw the object pull up sharply after passing the plane and disappear into a cloud. The sighting lasted only seconds, but in that flash of time the object made an indelible impression: "[It] appeared to be a wingless aircraft, 100 feet long, cigar shaped and about twice the diameter of a 'B-29,'" stated the report.

"It was clear there were no wings present, that it was powered by some jet or other type of power shooting flame from the rear some fifty feet," Captain Chiles said in a written statement to his employers at Eastern Air Lines. "There were two rows of windows, which indicated an upper and lower deck, from inside these windows a very bright light was glowing. Underneath the ship there was a blue glow of light. After it passed it pulled up into some light broken clouds and was lost from view. There was no prop wash or rough air felt as it passed."[1]

In John Whitted's statement, he estimated the speed of the object as 700 miles per hour, but added, "This is purely a rough estimate."

"After it passed us," Chiles later told a reporter, "we must have sat there for five minutes without saying a word, we were so speechless."[2]

Clarence McKelvie, an editor from Columbus, Ohio, was seated on the right side of the plane, in the fifth or sixth row of seats. He stated in his interview with a Project Sign investigator that he would periodically look out the window, and that the full moon made it easy to make out objects on the ground below. On one of those peeks, he saw "a sudden streak of light moving in a southern direction across the airway above the plane."

The streak appeared to be cherry red with a yellow edge. At first McKelvie thought it was lightning, but its straight-line movement told him that was impossible. He saw no physical shape—only a streak of flame, moving in a straight line.

If McKelvie was in any doubt about whether he had seen something unusual out his window, those doubts would have been erased by the pilots' behavior following the encounter: "Mr. McKelvie stated . . . that the pilots seemed quite excited and that they appeared nervous over the episode," the transcript concluded.

An hour before the rocket craft streaked past the DC-3, a ground crewman at Robins Air Force Base near Macon, Georgia, had noticed an unusual light approaching from the north, flying at approximately three thousand feet. "The first thing I saw was a stream of fire and I was undecided as to what it could be," stated Walter Massey when interviewed by a Project Sign investigator, "but as it got overhead, it was a fairly clear outline and appeared to be a cylindrical shaped object with a long stream of fire coming out of the tail end. I am sure it would not be a jet since I have observed P-84s in flight at night on two occasions."

By pure happenstance, Massey, only twenty-three years old, was one of the select few humans in 1948 to have actually seen both jet aircraft *and* rocketships in flight, giving him rather rare qualifications as an observer. "This was one of the fastest objects I have ever seen," he reported. "I saw German V-1s in the summer of 1944 and they were fast, but this one was even faster . . .

"During the Battle of the Bulge, a Sergeant and myself were on guard duty and saw something that resembled this object in question," he went on. "We later found that we had witnessed the launching of a German V-2 rocket. It carried a stream of fire that more or less resembled this object. This object looked like rocket propulsion rather than jet propulsion, but the speed and size was much greater."

Massey watched the flaming cylinder until it receded from view in the southwestern sky, on its way to its rendezvous with Eastern Air Lines Flight 576.

SIX MONTHS INTO PROJECT SIGN, the air force was caught flat-footed by the midair sighting. The Chiles-Whitted rocket, with its double-decker rows of windows and fifty-foot jet of red-orange flame, forced a paradigm shift upon the air force's investigators and the world.

Adding to the pressure on Project Sign was the fact that two of the four witnesses were all but unimpeachable. Both Chiles and Whitted had flown for the army in the war and had impeccable records. They were highly regarded by their employers at Eastern Air Lines. And they stuck with their story.

"According to the old timers at ATIC, this report shook them worse than the Mantell Incident," wrote air force intelligence officer Captain Edward Ruppelt. "This was the first time two reliable sources had been really close enough to anything resembling a UFO to get a good look and live to tell about it."[3]

Faced with screaming newspaper headlines describing the Chiles-Whitted rocket as a "Sky 'Devil-Ship,'" "Flame-Spitting Aerial Monster," "Mysterious Ball of Fire," and "Wingless Sky Monster," Project Sign had to give the public a reassuring explanation, and quickly. It was, after all, only six months since Thomas Mantell had crashed and died while chasing an unknown aerial object, and Project Sign's still-standing Venus explanation had failed to settle anyone's nerves. In an unprecedented move, Project Sign staff contacted every large civilian and military airfield in the southeast United States, as well as every commercial airline in the country, to ask if they had any ships on registered flight plans the night of July 24 that could have had a close flyby with the plane Captain Chiles was piloting. All but one of the 225 airfields and airlines that the air force reached out to responded in the negative, and the one positive—a C-47 military transport plane en route to an air base in Florida—did not match the description of the object's appearance or behavior in any way. In fact, the C-47 was a military twin of the DC-3 that Chiles and Whitted were flying; the two Eastern pilots would not be likely to misidentify a plane exactly like their own.

At first, all that an air force spokesperson could tell the public was that "'obviously' this country has no plane resembling a double-decked, jet-propelled, wingless transport shooting a 40-foot flame out of its back end."[4] At the same time, perhaps anticipating the air force's next move, Chiles and Whitted told an interviewer "they were certain it was not a meteor or a comet, because they had seen them before in the air."[5]

"There is no astronomical explanation, if we accept the report at face value," Hynek admitted in his lengthy analysis of the sighting, now known as Incident #144. "The sheer improbability of the facts as stated, particularly in the absence of any known aircraft in the vicinity, makes it necessary to see whether any other explanation, even though far-fetched, can be considered."

He then went on to consider just such a far-fetched explanation. Taking into account the "tremendous outburst of flame," the "cigar-shaped" fuselage, the "orange-red flame," the sighting duration of "five to ten seconds," and the fact that the object "disappeared into a cloud," Hynek wrote, "This much, at least, could be satisfied by a brilliant, slow-moving meteor." He felt that this hypothesis was supported by the fact that McKelvie, the passenger, had given a description that did not tally with that of a spaceship, "but does agree with that of a meteor."

"It will have to be left to the psychologists to tell us whether the immediate trail of a bright meteor could produce the subjective impression of a ship with lighted windows," Hynek wrote.

Massey's observation of a similar object from the ground an hour earlier (now known as Incident #144b) proved to be somewhat vexing to Hynek: "If those two sightings refer to the same object, there are two possible interpretations," he wrote. "One is that the object was some sort of aircraft, regardless of its bizarre nature. The other possible explanation is that the object was a fireball."

Massey's sighting had reportedly taken place an hour earlier than the Chiles-Whitted sighting and two hundred miles to the east-northeast, and both Massey and Whitted had estimated the object's speed at 700 miles per hour. At that speed, the object could have covered the distance of two hundred miles in about twenty minutes, assuming it never changed course or changed speed. A conventional aircraft of the time could cover the two hundred miles in an hour.

To further complicate the scenario, because the object was traveling east to west, it crossed over from the Eastern time zone to the Central time zone sometime between the two sightings, thus "gaining" an hour on the witnesses. Furthermore, any one of the witnesses could have failed to factor in the one-hour difference of daylight saving time, which had just been reinstated during World War II and thus was apt to cause miscalculations among even the most orderly minds. As a result, depending on how one looked at it, the two sightings could have taken place one hour apart, two hours apart, or instantaneously. Or there could have been two rockets.

The key was whether Massey, Chiles, or Whitted had been mistaken about the time of their sightings. Hynek decided that if the reported times were accurate, and if both times were reported as *Eastern* time, the object was an airplane, but if the times were incorrect the object was a meteor. Just three and a half pages after claiming that there was no astronomical explanation, Hynek concluded, "The object reported in incident #144a and #144b was very probably a meteor."

But his assessment was ignored—and the incident classified as "unknown"—because the winds were changing in the Project Sign offices. An agenda was taking shape, and mundane explanations like "very probably a meteor" did not fit the new narrative.

Project Sign didn't know what it had in the Chiles-Whitted Incident, but those on staff who were on the side of the extraterrestrial hypothesis took advantage of this moment of uncertainty. "They wrote an unofficial 'Estimate of the Situation,' classified top secret . . . it concluded that the evidence indicated the UFOs were of extraterrestrial origin," Jacobs reported.[6]

Submitted in the autumn of 1948, the "Estimate" worked its way up the chain of command without encountering any significant resistance, suggesting either that a great many people in charge at the air force were supportive of the extraterrestrial hypothesis or that nobody wanted to take a stand one way or the other.

The Estimate made it all the way up to the desk of air force chief of staff General Hoyt S. Vandenberg, where it met its doom. "The general wouldn't buy interplanetary vehicles. The report lacked proof," Ruppelt wrote. "A group from ATIC went to the Pentagon to bolster their position but had no luck, the Chief of Staff just couldn't be convinced."[7]

According to Ruppelt, the Estimate was "declassified and relegated to the incinerator. A few copies, one of which I saw, were kept as mementos of the golden days of UFO's."[8]

"This turn of events caused proponents of interplanetary visitation to go out of official favor and be reassigned," reported UFO historian Jerome Clark. "Officers who considered UFOs to be misinterpreted mundane phenomena rose to power."[9]

Hynek, meanwhile, was back in Columbus, teaching astronomy to eager OSU students by day, measuring the spectra of binary stars by night, adding a second son, Joel, to his family, and continually branching out in new directions. As he struggled to build up enrollment in the astronomy department (there were still only one or two astronomy majors per year), he was proving himself to be a formidable fund-raiser, securing "a gift to the department of physics and astronomy from alumni and friends of the University"[10] in the amount of $720 to pay for a new planetarium.

The planetarium became the first of its kind in Ohio and in many ways rivaled larger planetariums, such as the Adler in Chicago and the Hayden in New York City. "[Hynek] says, 'there is no more effective means of teaching elementary astronomy than by using a planetarium,'" reported the *Ohio State University Monthly*. "His students would undoubtedly agree enthusiastically, adding that the new equipment has brought great new interest to their study."[11]

Hynek continued to publish in scholarly journals, but his by-line was now appearing in more mainstream publications, such as *Nature, Practical Astronomy,* and *Sky & Telescope* magazine. He was also tapped by McGraw-Hill Publishing to edit, and write a chapter for, a new textbook simply named *Astrophysics.* The book, the first to bear Hynek's name, was published in 1951, some thirty-three years after he had his first look at an astronomy textbook as a child.

HYNEK'S TEACHING POSITION OFFERED AN UNEXPECTED PERK: since the 1920s the lakes of Ontario had been a popular summer destination for Ohio Wesleyan faculty, and when property on favorite fishing spot Lake Duborne became available, a group of professors purchased the land and sold it off in lots to OWU faculty, thus forming the Blind River Association. Hynek purchased a parcel on the lake and built the family cabin on Battle Point in 1947.

"It was a strong fixture in our family, and still is," said Paul Hynek. "I've been there forty-five times, I guess. It's an absolutely wonderful place; it's a cabin on a lake, no running water, there's no road—you have to take a boat to it—there's no electricity. I remember really well some nights seeing the northern lights as they covered the entire sky."

Adding to the charm, Hynek had a snug ten-by-ten-foot cottage—"a study in the woods"—built behind the main cabin. Here he would do most of his writing and, according to Paul, "get away" from the kids. "My mom would be out in the rain beating the cloth diapers against the rocks," he recalled, "while my dad was in his study with his little fireplace, cooking a hot dog and writing." The sense of isolation and escape was enhanced by the addition of a battery-powered shortwave radio, which Hynek would tune in to Radio Moscow for hours on end. "Radio was magic to him," said Paul.

It was by now clear to everyone involved that Project Sign had never come close to cracking the case of the flying discs. In early 1949, just a year after the project began, it was officially and unceremoniously canceled. Hynek's contract expired that April, and Captain Sneider and his staff were quietly reassigned. The case files were left to collect dust, and the extraterrestrial theory was shown the door.

The problem was that dozens of new saucer reports came in every month, and the air force still needed a place for those reports to go and die. Thus was born Project Grudge, a toothless investigative unit with a minuscule staff and budget left over from Project Sign. Project Grudge slipped right into the old Sign offices at Wright-Patterson and went straight to work debunking reports as quickly as it could.

"Everything was being evaluated on the premise that UFO's couldn't exist. No matter what you see or hear, don't believe it . . ." wrote Ruppelt. "With the new name and the new personnel came the new objective, get rid of the UFO's."[12]

By the time Hynek completed and submitted his final Project Sign report, Sign had given way to Grudge (leading many to believe incorrectly that Hynek had been contracted to work on Grudge). To fit the objectives of the new program, Hynek's report on the Chiles-Whitted object was revised a second time without his knowledge. The "wingless sky monster" that had terrified the nation was reclassified as "an obvious fireball."

While Project Grudge retained the classified status of Project Sign, it was important that the public have a clear idea of Grudge's mission. Articles placed in popular magazines portrayed flying saucer sighting reports as pranks, mistakes, and delusions and reassured the public that there was nothing to fear. The more the air force turned to the mass media to convince Americans that investigating flying saucer reports was a waste of time, however, the more reports it would receive and the less confidence the public had in the air force's credibility on the subject.

The staff of Project Grudge took only eight months to investigate 244 UFO reports and come to the unsurprising conclusion that flying saucers constitute no threat to the security of the United States. "Almost every incident has less than legal proof that an object was seen or that an object appeared or performed as described," read the Project Grudge final report. "Even in cases where more than one witness observed the incident, all witnesses seldom agreed on details." Further thinning the ranks of reliable witnesses, Project Grudge investigators rejected some because they had "too vivid imaginations, were of low intelligence, and were of questionable character."

After dismissing every case as a mistake, a hoax, or the product of a wild, perhaps unhinged imagination, Project Grudge recommended to the Pentagon that the investigation and study of unidentified flying objects should be "reduced in scope," and that new standards be developed so that the air force would have to deal only with reports "clearly indicating realistic technical applications."

The program's transparent directive to ignore the reality of the situation inspired derision and accomplished something completely unexpected and unheard of: it created a burgeoning market for flying saucer news and entertainment.

In sunny Hollywood, a little-known writer who normally reported on show business news took aim at the flying saucer scare, captivating the American public with a fantastic story of alien creatures from another world. *Variety* columnist Frank Scully debuted his sensational book, *Behind the Flying Saucers,* in 1950 and became the second Scully to be embroiled in a hoax involving a mysterious aerial phenomenon. The centerpiece of the bestselling book was the remarkable claim that the U.S. government had in 1947 and 1948 salvaged not one but three disabled flying saucers in Arizona and New Mexico—a fourth saucer was nearly nabbed but got away in the nick of time—and with them the remains of thirty-four unusual pilots and crew.

Scully credited this story to two mysterious sources, a Dr. Gee and his colleague Silas Newton, who claimed to have examined the saucers and the aliens, none of which survived exposure to our atmosphere. The creatures were all reportedly between thirty-six and forty-two inches in height. They appeared to be quite human, although all had perfect teeth and were dressed in the style of the 1890s, a possible link with that era's mystery airship reports and the Aurora crash. The saucers were said to be both held together and propelled by subtle manipulations of gravity, and there was speculation that the discs and their occupants came from the planet Venus. Vehicles and crew were all reportedly spirited away to ATIC at Wright-Patterson Air Force Base, Hynek's erstwhile employer. Scully, "vouched for by his publisher,"[13] Henry Holt & Company, insisted that he had fully vetted his sources' scientific backgrounds and found their uncanny story to be true.

As copies of the book flew out the doors of bookstores across the country—*Behind the Flying Saucers* sold approximately sixty thousand copies and was widely reprinted—Scully's tale started to crumble. Dr. Gee and Silas Newton were revealed to be con men who traveled the West selling devices that they claimed could detect the presence of crude oil deposits in the ground. Scully, whether taken in by the con men or complicit in the fraud, never wrote about flying saucers again. But then, he never needed to.

CHAPTER 5

SCINTILLATIONS

WITH SO LITTLE BEING DONE officially to address the questions and anxieties of the public where UFOs were concerned, Hollywood sensed the country's mood and provided the necessary nightmare, in the shape of the 1951 film *The Thing from Another World*. Freely adapted from John W. Campbell's novella, "Who Goes There?", originally published in the pulp magazine *Astounding Science-Fiction*, *The Thing* terrified moviegoers across the country and single-handedly created the template for the alien invasion movie.

In the film, military men and scientists at an Arctic research base unwittingly free an intelligent alien monster from a saucer-shaped spaceship embedded in the ice. The Thing is revealed to be a humanoid vegetable that needs human blood to survive, and the humans must find a way to destroy it before it can reach civilization. Driven by curiosity to learn how the alien lives and reproduces, and anxious to communicate with a new life-form, the lead scientist endangers everyone at the station—and, by extension, the entire human race—by giving the creature free access to the base. When the military commander discovers this treachery and takes countermeasures, the scientist is reduced to the character whom audiences least want to see saved.

One particular moment from *The Thing* has become iconic

to flying saucer fans and, as we will see, influenced a celebrated filmmaker years later. When the Thing has been vanquished—the chastened scientist wounded but alive, alas—the newspaper reporter at the base radios a warning to the world: "Watch the skies, everywhere!" he warns both his listeners and, none too subtly, the movie audience. "Keep looking! Keep watching the skies!"

Watch they did, and then they reported what they saw.

DONALD KEYHOE, a retired U.S. Navy major and former marine, had turned to writing when an injury hampered his ability to serve on active duty. Commissioned by *True* magazine to write a feature on the flying saucer phenomenon, Keyhoe surveyed the wreckage of Project Sign and the black hole of Project Grudge and came away convinced that the air force was concealing proof that Earth was being visited by extraterrestrial life-forms. How else could he explain why no one at the air force would take his calls?

Keyhoe's paranoid shocker, "Flying Saucers Are Real," appeared in the January 1950 issue of *True* and found such an eager audience that the publisher could barely keep up with demand at the nation's newsstands. "It was one of the most widely read and discussed articles in publishing history," wrote David Jacobs,[1] and it was quickly extended to book length. When the book version came out, it became a smash bestseller, moving more than five hundred thousand copies.

Like Frank Scully's earlier book, Keyhoe's missive was criticized for its wobbly research, reliance on anonymous interviews, and "a large amount of loose thinking,"[2] but unlike Scully, Keyhoe could never be accused of presenting fraudulent information as fact. He merely strung together snippets of information from a variety of sources into a persuasive stew of theory and conjecture. The Mantell crash? Obviously a fatal encounter with a spacecraft. The Chiles-Whitted case? Clearly a flyby of an alien rocket. Because the air force chose not to respond to Keyhoe's assertions, the allegations could never exactly be proven wrong, and the cover-up narrative started to accumulate a certain weight.

BLISSFULLY DISCONNECTED FROM THE AIR FORCE'S UFO WOES, Hynek continued to define himself in the world of academia. Now a full professor and assistant dean of the graduate school, he was a rising star at OSU and the university's first bona fide television personality.

While waiting for its broadcasting license to be approved by the FCC and for funding to be appropriated by the state legislature for its own campus TV studio, OSU debuted a ten-part Sunday evening television series on WLW-C, a commercial station in Columbus. *It's Your World* was designed to showcase the work of OSU's science faculty, and the first installment introduced viewers to a soft-spoken, bespectacled astronomy professor who described the wonders of the night sky.

"Hynek illustrated his TV presentation with picture slides showing galaxies and other views of the heavens through the world's largest telescope," read an article in the May 1951 *Ohio State University Monthly*. "Another very interesting part of the program was his demonstration of a planetarium, which reproduced the planetary system in miniature for teaching-lecture purposes."[3]

Hynek's popularity and effectiveness as an instructor, meanwhile, were growing by leaps and bounds. A feature story in the July 1952 issue of the *Ohio State University Monthly* told of Ohio's "top high school scholar" hitchhiking across the state to meet Hynek and enroll as an astronomy major. The student, Jack Wright from Warnock, Ohio, had chosen the astronomy program at OSU "after careful consideration of many schools," the article claimed. "He was most impressed by the top standing of the department of physics and astronomy. After his visit to the campus he felt he had made a wise choice."[4]

"Student interest in astronomy at Ohio State has enjoyed an upsurge in the past few years," the article went on to say. "Much of the credit is due to Dr. Hynek, who started a non-math astronomy course in 1947. His enthusiasm and grasp of the subject have kept enrollments in the elective course increasing each quarter."[5]

Indeed, Astronomy 500 was "just about always maxed out,"

according to Jennie Zeidman (née Gluck), who took the class in the fall quarter of 1952. It was a tremendously popular class for students who needed to fulfill a physical sciences requirement for their degrees, but it was also entertaining: "Hynek came in the first day and wrote his name on the board. 'My name is Hynek,' he said, 'as in giraffe.'"[6]

Zeidman recalled that Hynek was very open in class about his consultancy work for the air force's UFO project and in fact used his UFO work as a teaching tool of sorts. "Since World War II, people had started looking up at the sky more, and they were noticing things—perfectly normal things—that they just had never noticed before. By the completion of Astronomy 500, Hynek said, he hoped we would all be educated enough so that we would never feel the need to report a flying saucer," Zeidman recalled.[7]

Yet on a field trip to the Perkins Observatory that fall, Hynek and the entire awestruck class saw an unidentifiable object, an ellipse of bright lights, moving slowly across the sky near the horizon. Zeidman, a private pilot, identified the object the next day as a KC-97 Stratotanker returning from a night refueling mission, and immediately distinguished herself in class.

That auspicious moment marked the beginning of a long and vital friendship, but one in which Zeidman would often be forced to divine the true nature of Hynek's activities on her own, even decades after she became his trusted research assistant.

Despite the cancellation of its flying saucer project, the U.S. Air Force continued to knock on Hynek's door regularly. It seemed there was no end to the research the air force wanted Hynek to participate in, so he continued to sign on to projects on behalf of the Ohio State University Research Foundation throughout the late 1940s and early 1950s.

One project in particular that kept the telescope at McMillin Observatory busy from 1952 to 1953 sounded as though it could have arisen from frustrated conversations about the Thomas Mantell case. Entitled "Fluctuations in Starlight and Skylight," this study sought to determine "the effect of the atmosphere on celestial images." In simple terms, it was a study of what makes stars twinkle. On a practical level, it was a study to help astronomers, military observers, and meteorologists make better identifications of scintillating celestial lights, as well as to tell the difference between scintillation and actual movement.

One phase of the study, for example, looked at the Gemini stars Castor and Pollux, to compare their patterns of scintillation. Castor is a binary star and Pollux is a single star; Hynek sought to determine whether they scintillated differently as a result, and indeed they did. Consequently, we now know that a binary star scintillates in its own unique manner, and therefore it may appear at first to be something other than a star, even to a trained observer.

Meanwhile, the January–March 1953 progress report on "Fluctuations in Starlight and Skylight" revealed that "Dr. Hynek conducted a joint colloquium at the Harvard College Observatory on the subject of 'Astronomical Seeing.' The meeting was well-attended by astronomers and physicists."

A theme in Dr. Hynek's pursuit of pure science was beginning to emerge.

CAPTAIN EDWARD J. RUPPELT was never supposed to get involved with flying saucers. A decorated World War II veteran and recent recipient of an aeronautical engineering degree, Ruppelt had been unexpectedly reactivated from the Air Force Reserve at the outset of the Korean War. He was assigned in January 1951 to ATIC at Wright-Patterson, where he went to work investigating the capabilities of the new Russian MiG-15 fighter that was being deployed in the Korean theater.

Ruppelt shared office space at ATIC with the staff of Project Grudge, and he soon became curious about the office politics involved in Grudge's work. "I had been at ATIC only eight and a half hours when I first heard the words 'flying saucer' officially used . . ." he wrote. "When I came to work on my second morning at ATIC and heard the words 'flying saucer report' being talked about and saw a group of people standing around the chief of the UFO project's desk I about sprung an eardrum listening to what they had to say. It seemed to be a big deal—except that most of them were laughing."[8]

Ruppelt assumed that the men were laughing about a hoax or hallucination, but the case in question turned out to be a very credible sighting involving airline pilots and air traffic controllers at the Sioux City, Iowa, airport. He was shocked by the mocking laughter and started to pay closer attention to the flying saucer talk a few desks down.

"The one thing that stood out to me, being unindoctrinated in the ways of UFO lore, was the schizophrenic approach so many people at ATIC took," he wrote. "On the surface they sided with the belly-laughers on any saucer issue, but if you were alone with them and started to ridicule the subject, they defended it or at least took an active interest."[9]

What Ruppelt soon learned was that the air force's UFO activities were in complete disarray. The lurch from extraterrestrial-oriented Project Sign to debunking-oriented Project Grudge had caused great confusion within the air force, had created apathy tinged with hostility in the press, and had fostered enduring mistrust in the American public. And no one seemed to know how to fix the problem.

The air force hadn't even succeeded in shutting down Project Grudge. "From the beginning of 1950 until the middle of 1951 Project Grudge remained in a state of suspended animation . . ." wrote Jacobs. "By the summer of 1951 Project Grudge had so drastically reduced its staff that only one person, a lieutenant, served as investigator."[10]

Then on September 10 of that year, a sensational sighting over the Army Signal Corps radar center at Fort Monmouth, New Jersey, awakened the sleeping giant. An army pilot and his passenger, an air force major, spotted a disc-shaped object thirty to fifty feet in diameter that hovered in the air and evaded their attempts to approach it. On the ground, meanwhile, a radar operator was giving a demonstration to a group of officers when he tracked an object overhead going as fast as 700 miles an hour.

The director of air force intelligence, Major General C. P. Cabell, ordered a full investigation by Project Grudge, followed by a briefing at the Pentagon. The investigator, Lieutenant Jerry Cummings, spent a few days at Fort Monmouth and then reported to General Cabell that the object was a) a balloon, and b) a freak radar return "caused by unusual atmospheric conditions."[11] This dismissive report—which, one must remember, was entirely in keeping with Project Grudge's official directives—did not go over well, and the general asked for a full briefing on the status of Project Grudge. When Cabell learned that the project was essentially dormant, he hit the ceiling. "Who in hell," he asked Cummings, "has been giving me these reports that every decent flying saucer sighting is being investigated?"[12]

General Cabell authorized the reactivation of Project Grudge on the spot, but as Cummings was already scheduled to return to civilian life, the job fell to an ATIC officer who had already distinguished himself as someone who could straighten out "fouled-up projects"[13]: Captain Edward Ruppelt.

Given carte blanche to reinvigorate the program, Ruppelt set about rethinking and redesigning the way flying saucer cases were reported and investigated. But he did far more than completely revamp the operational aspects of the program; he also instituted a strict policy of unbiased reporting. Ruppelt's project would not promote an extraterrestrial theory or a terrestrial one, and any member of the staff who leaned too obviously in one direction or the other would be quickly reassigned. The saucers were given a less evocative name as well: they were now officially referred to as "Unidentified Flying Objects," or UFOs, after an army staffer inadvertently coined the term while investigating a 1947 New Mexico sighting.

Within months, Ruppelt had Project Grudge running so effectively that he started to get additional staff and resources, and the project was given a new moniker: Blue Book. "For those people who like to try to read a hidden meaning into a name, I'll say that the code name Blue Book was derived from the title given to college tests," said Ruppelt. "Both the tests and the project had an abundance of equally confusing questions."[14]

It was early 1952, and UFOs were respectable once more. New cases were being investigated with more care and old cases were being reconsidered. Among them was the Mantell case.

When the Air Force Office of Public Information started to receive new requests for information about Captain Mantell's 1948 crash, Ruppelt reopened the file and found that one of the initial investigators was still in the area. He called Dr. Hynek at Ohio State and set up a meeting for the very next day.

"Dr. Hynek was one of the most impressive scientists I met while working on the UFO project, and I met a good many," Ruppelt recalled. "He didn't do two things that some of them did: give you the answer before he knew the question; or immediately begin to expound on his accomplishments in the field of science."[15]

In his report for Project Sign nearly three years earlier, Hynek had declared that Captain Mantell had died while chasing the planet Venus. But Ruppelt wanted to review the entire incident—much of the original material in the file at ATIC had been damaged by a catastrophic coffee spill and was no longer legible—and Hynek didn't hesitate to retrieve his own notes on the case. He looked over his records and then did a truly remarkable thing: he recanted.

"[Hynek] had been responsible for the weasel-worded report that the Air Force released in late 1949," Ruppelt wrote, "and he apologized for it."[16]

What had changed since Hynek had declared that Captain Mantell had been chasing the planet Venus? The answer is really quite simple: month after month, year after year, *the UFO phenomenon persisted.*

"For several years I was saying there was nothing to it . . ." Hynek explained. "I thought the whole thing was a fad, a craze—and would pass from the scene as fads invariably do. Back in 1948, when I first started, I would have taken just about any bet that by 1952 the whole matter would be forgotten. It was the persistence of the phenomenon, not only in the United States, but over the world, that finally grabbed my attention."[17]

One of Hynek's primary assumptions about the flying saucers

was shattered by a single fact. If one assumption had been erroneous, couldn't they all be?

Ruppelt's research had shown that Hynek was correct about the azimuth and elevation of Venus the afternoon of the Mantell Incident, but the planet's brightness was an issue. "[Hynek] had computed the brilliance of the planet, and on the day in question it was only six times as bright as the surrounding sky," Ruppelt wrote, explaining, "Six times may sound like a lot, but it isn't. When you start looking for a pinpoint of light only six times as bright as the surrounding sky, it's almost impossible to find it, even on a clear day."

To Ruppelt's surprise, Hynek admitted to an error: he didn't think that the Mantell UFO was Venus after all.

What must it have felt like for Hynek to change his verdict on such a bellwether case? He never seems to have discussed the reasons for his change of heart with any of his colleagues, nor does he seem to have addressed it in any of his speeches, books, or articles. But his close associate Jennie Zeidman described Hynek to attendees at a UFO conference in November 1999 in revealing terms: "Over the total of 33 years that I worked with Hynek, I would say the prominent quality he evinced was intellectual curiosity—the search for knowledge—for answers—for their own sake—irregardless [sic] of what those answers might be . . . assumptions, wishful thinking, beliefs, and theories may have absolutely no relationships to the FACTS."[18]

Captain Ruppelt left OSU feeling quite impressed with the Project Sign astronomer, and determined to find a better explanation for the Mantell case. His inquiries led him to the U.S. Navy Office of Naval Research, which operated a classified high-altitude atmospheric research program called Project Skyhook, utilizing silver balloons that could expand to as much as thirty meters in diameter once aloft. The first Skyhook balloons were secretly launched in late 1947, several months before the Mantell crash, but the launches were still need-to-know in January 1948, when Captain Mantell had his aerial encounter. Was it possible, Ruppelt wondered, that the object that had been sighted soaring across the skies of northern Kentucky that afternoon had been a thirty-meter silver Skyhook balloon whose existence had been kept under wraps?

Ruppelt found that the balloon theory had been considered by the Project Sign investigators, but it had been rejected for two reasons: "Number one was that everybody at ATIC was convinced that the object Mantell was after was a spaceship and that this was the only course they had pursued. When the sighting grew older and no spaceship proof could be found, everybody jumped on the Venus bandwagon, as this theory had 'already been established.' It was an easy way out. The second reason was that a quick check had been made on weather balloons and none were in the area."[19]

By 1952, Project Skyhook had been declassified, however, and Ruppelt discovered that some early launches had taken place at Clinton County Air Force Base in southern Ohio. He consulted the air force's weather records from the day of the Mantell crash and found that the prevailing winds could have swept a large balloon launched from Clinton County over the areas from which the strange object had been sighted. With the planet Venus out of the picture and Project Skyhook suddenly looming large, the Mantell Incident took on a whole new configuration.

"It *could* have been a balloon," Ruppelt concluded. "This is the answer I phoned back to the Pentagon."

Up until the summer of 1952, UFOs had, with few exceptions, limited their appearances to the empty skies above remote, scarcely populated areas. That July, they moved to the city—Washington, D.C., to be precise.

The first sighting of what would come to be known as the "Washington Merry-Go-Round" took place just before midnight on Saturday, July 19, when air traffic controllers at Washington National Airport (now Ronald Reagan Washington National Airport) spotted multiple fast-moving objects on radar. The readings were confirmed by both the long- and short-range radars at the airport and by control tower personnel at nearby Andrews Air Force Base.

Throughout the night of the nineteenth and into the early-morning hours of the twentieth, numerous unusual objects—from bright, flaming orbs to metallic discs to a huge, fiery orange sphere—were spotted in the sky and picked up by radars all across the region. The credibility factor was unusually high, as many of the visual sightings were made by control tower personnel and commercial pilots.

Even with such an embarrassment of riches—the objects were appearing in groups of up to a dozen at a time—the air force proved itself completely inept at tracking and identifying the objects, which seemed to disappear from both the sky and radar screens at will, then reappear somewhere else almost instantaneously. The chief air traffic controller at Washington National, Harry Barnes, found himself at a loss when writing up a report for his employers at the Civil Aeronautics Administration: "It would be extremely difficult to write this so that it is in a logical sequence due to the confusion that seems to have existed throughout the entire affair."[20]

Barnes spent much of the night on the phone with personnel at Andrews Air Force Base, trying to determine who was in charge and what was being done to identify the objects. He was told at various times that the air force was "doing nothing about it," that "all information is being forwarded to a higher authority," that the air force "would not discuss it any further," that "somebody else was supposed to handle it," and that the base intelligence officer "had gone back to bed and the report would go in later."[21]

Later, Barnes described the uncanny perception he had of the radar targets, or "pips," he tracked for six hours that night: "The only recognizable behavior pattern which occurred to me from watching the pips was that they acted like a bunch of small kids out playing. It was helter-skelter, as if directed by some innate curiosity. At times they moved as a group or cluster. Other times as individuals over widely scattered areas."[22] Barnes went on to state unequivocally that the unidentifiable objects in the sky over Washington, D.C., that night "were not ordinary aircraft."[23]

The following weekend it all started again, and this time the confusion factor went off the charts. "[Washington National] and

Andrews were tracking more targets than they could handle," reported Jerome Clark. "Sometimes they moved more slowly, at less than 100 miles per hour, and sometimes abruptly reversed direction and streaked across the sky at what calculations indicated was 7000 mph."[24]

Officers at Andrews found the radar returns solid enough that they scrambled planes to intercept the objects. But as the jets approached the UFOs, the targets simply disappeared, leading one pilot to remark on the "incredible speed of the object."[25]

Air force spokesperson Al Chop recalled that the jets circled around for fifteen minutes after the objects disappeared from radar, but their reconnaissance came up negative. Chop, who had been roused out of bed when the objects started to appear that night, was in the control tower at Andrews the whole time and saw what happened when the jets were ordered back to base: "As our interceptors left the scope, our targets reappeared."[26]

Fed by the chaos and uncertainty of the incident, public interest in UFOs spiked—and the scientific community struggled to respond to the country's need for reassurance. A July 28 wire service story made it clear that, when it came to flying saucers, the smartest people in the nation weren't sure of anything at all. One scientist who requested anonymity said that the "saucers" might be experimental aircraft developed by the United States, while an anonymous official at Mount Wilson Observatory in Los Angeles said, "The objects reportedly sighted near Washington apparently were not a form of meteor or 'other natural phenomenon.'"[27]

In the same article, Otto Struve, Hynek's old boss at Yerkes Observatory, said he was "baffled" by the reports. "I have no idea what flying saucers could possibly be," Struve declared. "I have no knowledge on the subject whatsoever."[28]

The real surprise in the article, however, came from Hynek, who "said he was convinced these persons saw something—'some type of object or phenomenon.'

"But, Dr. Hynek said, it is 'highly improbable' that the 'saucers' come from another planet."[29]

Hynek seemed to understand that in time the Washington

sightings would come to represent the very best type of UFO cases; visual sightings backed up by radar tracking are difficult to explain away, and in fact the Washington sightings—at least some of them—are still listed as "unknowns" to this day. But in 1952, the combination of visual and radar sightings simply caused massive confusion and inspired wholesale disbelief. Which is not to say that the Pentagon didn't realize what a problem it had on its hands. The fact that Andrews had scrambled jets to intercept the objects speaks to the gravity of the situation. And yet, when the air force announced at its hurried press conference that the sightings were the result of temperature inversions creating false radar echoes, reporters swallowed the story hook, line, and sinker.

Inexplicably, Project Blue Book was not called in to investigate the Washington Merry-Go-Round. Although Ruppelt was in Washington, D.C., at the time, he was kept away by a bizarre maze of restrictive Pentagon red tape—he was denied a staff car, denied money to arrange his own transportation, and then threatened with being reported as AWOL if he didn't return immediately to Wright-Patterson.

With UFOs in the headlines once again, Hynek set off on his very first clandestine mission, recruited by Columbus, Ohio–based Battelle Memorial Institute to secretly gather UFO intelligence on its behalf.

Ruppelt had recently hired Battelle to analyze the Soviet Union's "sneak attack" capabilities, but in reality the scope of its work went far higher than the stratosphere. Battelle was to conduct Project Stork, a statistical analysis of thousands of UFO reports using a machine indexing system, or what we now call a computer. What was it indexing? According to Hynek, Battelle was trying to answer no less than the essential UFO question: *What are they?*[30]

Ruppelt, for reasons of his own, dubbed Battelle Memorial Institute itself Project Bear and described its mission thusly: "Besides providing experts in every field of science, they would make two studies for us; a study of how much a person can be expected to see and remember from a UFO sighting, and a statistical study of UFO reports."[31]

To that end, Battelle would study all Sign, Grudge, and Blue Book reports through the end of 1952 to determine whether there were quantifiable differences between the "unknowns" and the "knowns." There was, however, another element of Battelle's research that could not be derived from the air force's case files. "They wanted," Hynek said, "to find out discreetly what astronomers felt about UFOs."[32]

The fact that Battelle offered him this project, almost certainly on Ruppelt's recommendation, may have surprised Hynek, who felt that the captain distrusted him.

"That was in my debunking days, and he regarded me with a certain amount of suspicion," Hynek recalled years later, adding, "He never really took me into his confidence . . ."[33]

To be fair to Ruppelt, however, Hynek may simply have been displacing his own distrust of his new project chief. After acknowledging that pleasing the Pentagon was "a most difficult task" for a Blue Book project chief, Hynek went on to write that Ruppelt "was there to tell the brass what UFOs were—not to perpetuate a mystery. Generals don't like mysteries; they want hard, crisp answers. 'We showed that it was a balloon' or 'it was definitely Venus' won more acclaim than 'we don't know what it is; it might be extraterrestrial but we are puzzled.'"[34]

On June 22, Project Stork sent Dr. Hynek out to rove around the country and on into Canada, visiting observatories and universities in an effort to sniff out professional astronomers who had actually seen UFOs but who had opted not to report their sightings to the air force.

Hynek set out to attend a meeting of the American Astronomical Society in Victoria, British Columbia, where he knew he would find many of the most important figures in the field. It was a unique and no doubt thrilling position in which Hynek found himself. He was, in some small part, helping to answer the question, "What are they?" but was able to surreptitiously sound out his peers and his elders in the astronomy field before staking out his own position on the matter.

"Whenever possible I brought up the subject in cocktail gatherings and in meetings," Hynek recalled. "I traveled down the West Coast, visiting Lick Observatory and Mount Wilson and so forth, and I tried my best cloak and dagger, completely unobtrusive manner . . ."[35]

Hynek visited eight observatories, polled forty-five astronomers, and found that 36 percent were "not interested at all or totally hostile to the subject," while 23 percent held that "UFOs represented a problem that was more serious than people recognized." A full 41 percent of the astronomers Hynek spoke with "were sufficiently interested in the whole subject to go so far as to offer their services if they ever were really needed." All in all, he noted in a December 1952 Project Blue Book status report, "actual hostility was rare."[36]

Five of the forty-five astronomers reported having seen UFOs themselves, but the unusually high percentage didn't surprise Hynek in the least. After all, he reasoned, astronomers spend more time than the average person watching the sky.

Over and over again on his journey, Hynek found that astronomers who joined in with their peers in publicly ridiculing UFOs would often become quite sympathetic when talking privately. He dubbed this the "committee complex" and described it thusly: "A scientist will confess in private to interest in a subject which is controversial or not scientifically acceptable but generally will not stand up and be counted when 'in committee.'"[37]

Hynek's determination to find a natural explanation for any and all UFO reports, already fading after Ruppelt asked him to reconsider his analysis of the Mantell crash, started to slip even more seriously as he discussed the topic with his colleagues. As he wrote in the Project Blue Book status report, "I took the time to talk rather seriously with a few of them, and to acquaint them with the fact that some of the sightings were truly puzzling and not at all easily explainable. Their interest was almost immediately aroused, indicating that their general lethargy is due to lack of information on the subject."

Based on his analysis of the forty-five interviews, Hynek called for UFO sighting reports to be treated as a scientific problem. Data from sightings should be "gathered with meticulous care and . . . weighed and considered, without rush, by entirely competent men." He conceded that these competent men might weigh the evidence and decide that the phenomenon wasn't worthy of study after all. "Personally, I hardly think that this will be the case," he wrote, "since the number of truly puzzling incidents is now impressive."

In his conclusion, Hynek all but admitted that he had been dead wrong about the phenomenon all along. "One final item," he wrote, "is that the flying-saucer sightings have not died down, as was confidently predicted some years ago when the first deluge of sightings was regarded as mass hysteria. Unless the problem is attacked scientifically, we can look forward to periodic recurrences of flying-saucer reports. It appears, indeed, that the flying saucer along with the automobile is here to stay."

CHAPTER 6

PROJECT HENRY

ENCLOSURE (1) IS FORWARDED HEREWITH FOR WHATEVER VALUE IT MAY HAVE IN CONNECTION WITH YOUR INVESTIGATION OF THE SO-CALLED "FLYING SAUCERS." PERTINENT DATA FOLLOWS . . .

SO BEGAN AN AUGUST 11, 1952, report sent by navy chief warrant officer Delbert Newhouse to Hill Air Force Base in Ogden, Utah. The pertinent data was "One (1) fifty-foot roll of processed 16mm color motion picture film," which would play a significant role in the investigation of "the so-called 'Flying Saucers.'" The fifty-foot magazine of Kodachrome Daylight movie film sent off to Utah that day had been exposed over a month earlier, on July 2, and resulted in forty seconds of motion picture film that even today defies explanation.

Chief Newhouse was on vacation with his wife, Norma, and their two teenaged children that day, driving south on U.S. Highway 30 several miles outside of Tremonton, Utah, at 11:10 A.M., when Norma commented on a group of unusual objects overhead. The sky that morning was a clear, cloudless blue, and Newhouse had no trouble spotting what his wife was looking at. He pulled the family car over to the side of the highway to get a better look.

"There was a group of about ten or twelve objects—that bore no relation to anything I had seen before—milling about in a rough formation and proceeding in a westerly direction," Newhouse wrote in his report. The objects were almost directly overhead at first, but in the time it took Newhouse to unpack the Bell & Howell movie camera and load the film magazine, the objects had moved a considerable distance away. Newhouse set the camera for an exposure level of f/8, set the focus on infinity, and started filming.

The images Newhouse captured are deceptively simple. A dozen or so bright white circular objects seem to move about randomly against a deep blue sky, but after several seconds their movements no longer seem quite so random. Some of the objects appear to group together in pairs, then form patterns, and then separate again only to reconverge. Their motion suddenly appears intentional, cyclical, controlled. A single object zips across the field of view away from the others. Then the image goes black.

"There was no reference point in the sky and it was impossible for me to make any estimate of speed, size, altitude or distance," he wrote. "Toward the end one of the objects reversed course and proceeded away from the main group. I held the camera still and allowed this single one to cross the field of view, picking it up again and repeating for three or four such passes. By this time all of the objects had disappeared."

Newhouse said the objects looked like "two saucers, one inverted on top of the other," and estimated their size to be equal to a B-29 at ten thousand feet. He described them as having a bright, silvery color and a metallic appearance. There was no doubt in Newhouse's mind that the objects were made of metal, and he had good reason to be so confident in his observational powers: not only was he a twenty-one-year navy veteran with some two thousand hours of flight time under his belt, but he also possessed a most curious and most appropriate job title: chief photographer.

NEWHOUSE'S FILM WAS ULTIMATELY SENT to the Air Force Photo Reconnaissance Laboratory, where the country's most talented photo analysts were charged with studying the luminosity, movement, and behavior of the objects in Newhouse's film.

Although Newhouse claimed that the footage "fell far short" of what he saw with his naked eye, his forty seconds of film baffled the intelligence experts at both the Air Force Photo Reconnaissance Laboratory and, later, the Naval Photographic Interpretation Center. Newhouse himself was described by the intelligence officer who interviewed him and his family as "quite possibly" an "expert photographer," and his footage was described in one Blue Book memo as "possibly the best documentary evidence yet obtained of unidentified flying objects" to have been analyzed by the air force's photo laboratory.

After an estimated one thousand man-hours of work, the U.S. military's finest photo analysts came to the following conclusions:

> *It is the majority opinion of the group conducting this analysis that these images are light sources. This will explain the non-blinking and variations in luminosity—but not the velocity or acceleration factors. In either case, light source or reflective surface, it appears as if the objects are of a nature which we are not able to identify in terms of natural phenomena or commonly known man-made objects.*
>
> *There is no indication of what kind of objects could have caused the images except that they must be of a construction, design, and material not commonly known. This is indicated by the computed acceleration rate and velocity . . . For the same reasons, birds, aircraft and balloons are ruled out.*

Years later, Arthur Lundahl was willing to go public with what he saw in 1952 as the director of the Naval Photographic Interpretation Center: "I've seen a genuine film of UFOs that, as a photo analyst, I believe could not have been faked."[1] Lundahl, whose other great contribution to U.S. intelligence was to confirm to President John F. Kennedy that there were, indeed, nuclear missiles in Cuba, was very clear about the Newhouse film. "We tested that film, about 1,600 frames of 16mm Kodachrome motion picture film, for several hundred man-hours on a frame projector. We analyzed each frame and looked for possible doctoring. We looked for flecks of dust that might have caused refracting, even for holes in the bellows of the camera. But we found nothing to explain what appeared on the film."

A press release issued jointly by the air force and navy made it clear that something out of the ordinary had occurred just outside of Tremonton. Citing a lack of detail in Newhouse's filmed images, however, the air force announced that it "will not speculate concerning the nature of the objects."

"The Navy analysts didn't use the words 'interplanetary spacecraft' when they told of their conclusions," Ruppelt wrote, "but they did say that the UFO's were intelligently controlled vehicles and that they weren't airplanes or birds."[2]

Clearly, Ruppelt wanted on some level for the navy analysts to use those words, for he felt he couldn't use them himself. "If I had to give an impression of [Ruppelt], it would be that he was sort of a weather vane," Hynek said. "He was extremely puzzled and one day he was in one direction and the next day his weather vane had shifted to the other. He was trying to do the best job he could to debunk it and yet he had this weird perception . . . something was going on that was beyond him."[3]

Something was indeed going on, and it was, in 1952, beyond Captain Ruppelt, beyond the air force, beyond Project Stork, and beyond the scientific establishment. The curious result of this institutional uncertainty was a growing desire to debunk and dismiss, and it reached notable heights at a gathering of the American Optical Society in Boston that October. What made the meeting so significant was that the society had invited three respected scientists to deliver research papers on the flying saucer phenomenon, and all three had accepted the offer.

The first speaker, Atomic Energy Commission member and Bendix Aviation researcher Dr. Urner Liddel, focused his wrath on the 1948 Chiles-Whitted case. Chiles and Whitted, Liddel explained, had fallen victim to an atmospheric illusion: "haze particles" in the clouds that night had created a mirror image of the DC-3, and the pilots had mistaken the phantom image for a solid physical object.[4]

Liddel's dismissal of the four-year-old case as a mere reflection of moonlight off a freak invisible cloud ignored much of the witnesses' testimony and turned what remained inside out. Undeterred, Liddel went on to reduce the American public to a horde of gullible fools who were so awed and enchanted by such recent scientific wonders as radio and television that they "will believe most any story,"[5] and declared that he saw no evidence that UFOs represented an extraterrestrial phenomenon.

A similar paper delivered at the meeting by Harvard astronomer Donald H. Menzel, an even more prominent figure in the American scientific establishment, dismissed all UFO sightings as "mirage, reflection, refraction, temperature inversion, and the like."[6]

As the acting director of the Harvard College Observatory in Cambridge, Massachusetts, Menzel had the highest profile of the three presenters and treated his appearance at the meeting as the opening salvo in what would become a long crusade to banish all consideration of UFOs from respectable scientific discourse. To Menzel, who had seen an unidentified flying object himself years earlier, all UFO reports were the result of misidentified, mundane atmospheric tics, and that was that.

Hynek was the third speaker in attendance that day, and he had some strong opinions about his colleagues' presentations.

"It was confirmed, as Dr. Hynek already believed," read the summation in Project Blue Book Report #9, "that Drs. Liddell [sic] and Menzel had not studied the literature and the evidence and, hence, were not qualified to speak with authority on the subject of recent sightings of unidentified aerial phenomena."

"Both papers," Hynek's report went on, "presented a series of well-worn statements as to how jet fighters, meteors, reflections from balloons and aircraft, and optical effects, such as sundogs and mirages, could give rise to 'flying saucer' reports. Since there was nothing new in either of the two papers, the trip from this standpoint was unproductive."

When it came time to give his own presentation, Hynek didn't just make the case for a more thorough investigation of the many "unknowns" in the Blue Book files—that would have been bold enough for a man in his position—but he was openly critical of the dismissive attitudes of his fellow speakers, Liddel and Menzel, two men who held positions of great power and influence in his world.

Hynek defined a flying saucer as "any aerial phenomenon or sighting that remains unexplained to the viewer at least long enough for him to write a report about it,"[7] and he abruptly skewered Menzel and Liddel for being too quick to seek out conventional explanations for UFOs and too willing to ignore evidence.

Then, like Clark Kent tearing off his tie and shirt to reveal his true heroic nature, Hynek boldly revealed his authority on the subject, by virtue of his involvement with the air force's flying saucer studies virtually from day one, and his possession of what was already becoming an encyclopedic knowledge of the phenomenon. He quickly laid out the parameters of what he considered a UFO report worthy of in-depth study: one that had not yet been explained, that had multiple witnesses, at least one of whom was a trained, practiced observer, and that lasted more than a minute.

After defending the air force's efforts to analyze, explain, and catalog the vast number of UFO reports it continued to receive, Hynek declared that these efforts must be directed toward a specific goal: if the as-yet-unexplained reports indicate a natural phenomenon, then that phenomenon must be rigorously studied; if they cannot be explained by natural causes, then there is an "obligation" to prove that they are balloons, mirages, or misinterpretations. He clearly felt that neither Menzel nor Liddel had met this standard.

"The chief point here," Hynek said, inching further out on his limb, "is to suggest that nothing constructive is accomplished for the public at large—and therefore for science in the long run by mere ridicule and the implication that sightings are the products of 'bird-brains' and 'intellectual flyweights.'

"Ridicule," he said, "is not a part of the scientific method and the public should not be taught that it is."[8]

FRESH OFF TWO MISSIONS FOR BATTELLE, Hynek was suddenly involved in the UFO mystery at a much higher level than ever before.

Most UFO reports, he had initially felt, involved misinterpretations of the actual facts. "But I also came to recognize," he added, "that there was a relatively small residue of UFO reports which were so well attested and so compellingly strange that the chances were overwhelmingly great that they could not be ascribed to collective misidentification, hoax, or hallucination."[9]

Suddenly, Hynek's 80 percent success rate at explaining away UFO reports was overshadowed by his inability to explain away the other 20 percent. The numbers hadn't changed; the man had.

Does this mean that Hynek regretted having dismissed so many witnesses' accounts of their compellingly strange experiences, and in so doing branding people like Kenneth Arnold, Clarence Chiles, and Thomas Mantell liars or fools? In one case, at least, the answer was yes.

On August 13, 1947, at about one P.M., A. C. Urie and his two sons saw a sky-blue metallic flying disc streaking down the Snake River Canyon about twenty miles from Twin Falls, Idaho. The object was flying about seventy-five feet above the floor of the four-hundred-foot-deep canyon, it emitted exhaust flames, and "as the machine went by the URIE place, the trees over which it almost directly passed (Morman Poplars) did not just bend with the wind as if a plane had gone by, but in URIE'S words, 'spun around on top as if they were in a vacuum.'"[10]

"Urie described the size as about 20 feet long by 10 feet high and 10 feet wide, giving it an oblong shape," read the report in the local newspaper. "It might be described as looking at an inverted pie-plate, or a broad-brimmed straw hat that had been compressed from two sides."[11]

Project Sign investigators found that Urie came across as sober and sincere. Not only did his two boys corroborate every detail of the description and behavior of the object, two other men, fishing nearby in the Snake that same afternoon, heard a roar and saw two strange aircraft high above them. One of the witnesses was County Commissioner L. W. Hawkins; the other wished to remain anonymous, as he had taken a sick day from work to spend the day recreating.

In the face of such compelling accounts from multiple witnesses, Hynek's Project Sign report was a masterpiece of illogic: "There is clearly nothing astronomical in this incident," he wrote. "Apparently it must be classified with the other bona fide disc sightings. Two points stand out, however: the 'sky blue' color and the fact that the trees 'spun around on top as if they were in a vacuum.' Could this, then, have been a rapidly traveling atmospheric eddy?"

"The Air Force was only too happy to accept my conjecture," Hynek explained. "'It seems logical,' the Project officer later confirmed, 'to concur with Dr. Hynek's deduction, that this object was simply a rapidly moving atmospheric eddy.'"[12]

There was one significant drawback to Hynek's deduction, however: "I had never seen an eddy like that," he admitted later, "and had no real reason to believe that one even existed."[13]

In his 1977 book, *The Hynek UFO Report,* he admitted that his concoction of a fanciful "atmospheric eddy" to dismiss the Urie sighting had been a grave mistake. "The fact that I have never seen such an 'eddy' (or as far as that goes, never even seen one described in books) and that I blithely discounted other pertinent evidence, haunts me to this day," he confessed. "I wonder what would have happened had I written: 'We must believe these witnesses, especially in view of the many similar reports received during the recent past. This was indeed a strange craft, involving technology far beyond ours.'"[14]

"In retrospect, I was a complete jerk," Hynek was to admit in a later interview.[15]

IT WASN'T LONG before the phenomenon attracted the kind of serious attention that Hynek now felt it deserved. Unaware that Project Blue Book had contracted with Battelle to analyze its backlog of UFO cases to look for patterns that separated the "knowns" from the "unknowns," influential parties at the Central Intelligence Agency (CIA) decided to take a close look at the case files and perform their own threat analysis.

In the wake of the confusion and consternation stirred up by the Washington Merry-Go-Round, the CIA Office of Scientific Intelligence was alarmed that defense capabilities in the airspace above the nation's capital could be so easily tied in knots. "On the two weekends when sightings were at their most intense, intelligence channels were clogged with UFO-related communications," wrote Jerome Clark. "Air Force generals and CIA officials worried that an earthly enemy such as the Soviet Union could take advantage of such a logjam by launching an attack if it so chose. Or it could use the UFO reports to confuse Americans and undermine confidence in their leaders."[16]

The result of the agency's brief but intense interest in the UFO phenomenon was the creation of the Robertson Panel, chaired by Dr. H. P. Robertson, a classified CIA staffer and director of the Defense Department's Weapons Systems Evaluation Group (that title alone is a clear indication of the panel's objective: evaluating what the CIA feared was a new type of weapons system). Robertson and a small group of cream-of-the-crop scientists convened in January 1953 to conduct what was lauded as the definitive scientific analysis of the UFO problem.

Battelle strongly objected to the formation of the panel, protesting that its own far-reaching Project Stork would render a much fuller and more reliable analysis of the UFO phenomenon, given time. But national security could not wait, and at the CIA's insistence the panel met in Washington, D.C., on January 14, 1953, to begin work on this most perplexing mystery. Both Captain Ruppelt and Dr. Hynek were enlisted as associate members of the panel, but Hynek quickly discovered that in this group, "associate" was a nice word for "outsider."

"Certainly, throughout the whole meeting, I did not feel at all like a colleague of the panel," Hynek recalled, "but rather as one of the witnesses brought in for certain evidence or comments, and then dismissed as a witness would be when he's asked to step down from the chair."[17]

Brought in on the second day of what would prove to be a four-day sprint through five years of accumulated data, Hynek and Ruppelt joined the scientists in a screening of Delbert Newhouse's film and another, similarly controversial movie taken earlier in Great Falls, Montana. This was a tremendous moment; these films had come to represent what "some intelligence officers considered as the 'positive proof'" that UFOs were real physical objects, according to Ruppelt.[18] But Hynek's enthusiasm waned when he realized the films would be projected not onto a movie screen but onto the bare wall of the conference room, hardly the type of viewing environment in which anyone could make a valid judgment of the authenticity of the images.

In fact, the entire event was marked by shoddy planning and lazy execution. The panel considered only 23 UFO reports out of the 4,400 that had been received by ATIC since 1947. One of the panelists did not arrive until the third day of the gathering, but his opinions were accorded equal weight to those of the others. A movie of a flock of seagulls was shown as an illustration that the objects in the Tremonton film might well have been birds, but it was shown a full day later, making reasoned comparisons all but impossible.

Not that that troubled the panelists in the slightest. "If the whole Robertson Panel was a put-up job," Hynek reasoned, "then one could argue that they deliberately chose high scientific-establishment men, men who were terribly terribly busy, could obviously not spend a lot of time examining things and had no intention of doing their homework."[19]

So inhospitable was the environment within the Robertson Panel, Hynek decided not to share key evidence that would have challenged the scientists' biases. "Do I have any outstanding evidence to present to the contrary? Well, I thought I might have had," he admitted, "but in the face of that onslaught, I wasn't about to bring it up; and furthermore, the data at that time was not convincing in the sense that a physicist wants something to be convincing . . ."[20]

The loss of another opportunity was tangible. The panel found that UFOs themselves did not pose a threat to the security of the United States, but that public interest in UFOs—and the public's growing tendency to report UFO sightings—was of grave concern. All other considerations were swept aside, and with them Hynek's illusions.

"Another way to describe their basic attitude," said Hynek of the Robertson panelists, "was very clearly an attitude of 'Daddy knows best, don't come to me with these silly stories, I know what's good for you and don't argue.'"[21]

To that end, the panel recommended to the CIA and the air force that every effort should be made to make the public lose interest in UFOs. It was an interesting and original twist on the Project Grudge mandate to "make the UFOs go away," but equally nonsensical. Paradoxically, the panel's recommendation to minimize the problem meant beefing up Project Blue Book with additional staff and resources, so that Captain Ruppelt's team could investigate—and then explain away—more UFO reports.

Thus did the newly enlightened Hynek—elevated suddenly to a bona fide UFO investigator and, just as suddenly, driven by a vital interest in investigating UFOs—join the staff of the newly expanded Project Blue Book, to reassure the public that the phenomenon that had so recently overwhelmed the country's air defense systems and convinced the country's intelligence services of its danger was nothing but a bogeyman. By this point Hynek must have felt that he had surely gone down the rabbit hole.

In less than a month, Delbert Newhouse's movie film had been downgraded from the most convincing evidence of the physical reality of UFOs ever examined by the U.S. government to a meaningless recording of misidentified seagulls, or perhaps "pillow balloons" launched as part of a propaganda campaign to support Radio Free Europe. This despite the fact that those particular balloons were not launched in the United States until February 1954, a full nineteen months after the Tremonton film was shot.

The final Blue Book notation on the Tremonton sighting stated the following: "Poss Explanation Pillow Balloons or Birds." But in a bold new development in the air force's treatment of UFO reports, Blue Book files show that the abbreviation "Poss" for "Possible" has been scribbled out and handwritten over with "Prob" for "Probable."

WHEN HYNEK RETURNED TO OSU from his frustrating experience with the Robertson Panel, he had a surprising new ally and confidante: Jennie Zeidman, the enterprising student who had solved the mystery of the UFO sighted by Hynek's astronomy class earlier in the school year. She had taken on a research assistant position with the doctor shortly before he left for Washington, D.C.

Zeidman remembered Hynek's first day back after the Robertson Panel meeting as fittingly cold and wintry. "I expected him to announce there would be a major scientific undertaking on the subject," she wrote in *International UFO Reporter*. "Instead, he told me, 'They're not going to have a scientific investigation. For some strange reason they voted it down. They didn't even take a decent look at the data, and they decided to discredit them.'"[22]

To her amazement, Hynek seemed to have already shaken off his disappointment and was starting to consider his newly expanded role in Project Blue Book.

"Perhaps he needed some levity that day when he looked up from his coffee and crackers and suggested that we should have a name for his consultancy project, 'something that captures the idea that these things flit around the sky,'" Zeidman wrote.[23] Recalling a brand of insecticide called Flit, and its popular advertisements in which a woman spots a bug in the house and cries out to her husband, *"Quick, Henry, the Flit!"* Hynek came up with a suitably oddball new code name for his UFO consultancy work: Project Henry.

In reality, the code name had a certain logic to it: the Czech surname "Hynek" is a variant of the German "Heinrich," which in turn is another form of "Henry."

Thus, in the early months of 1953, Hynek and Zeidman—

herself newly cleared for top secret work at age twenty-two—began sending regular reports on the status of Project Henry to their bosses at Project Stork, who, under the moniker Project Bear, would in turn forward them to Captain Ruppelt at Project Blue Book.

"Readers familiar with Hynek's sense of humor know I couldn't possibly have made this up," Zeidman wrote.[24]

Hynek and Zeidman proved to be increasingly efficient at their UFO work and established an extremely productive reporting system to tie Henry, Stork, Bear, and Blue Book together. "About once a week a courier . . . would arrive at Hynek's office at Ohio State University with a manila envelope stuffed with TWXs—teletype UFO sighting reports received from military facilities around the world," Zeidman wrote.[25] Hynek would then go over the new reports and do his best to determine whether any of the UFO sightings could be attributed to astronomical objects or events. Strictly speaking, this was all that was required of Hynek, but from week to week and from month to month, he quietly expanded his role and his reach.

"When we would get a report of high strangeness, he would scratch his chin—beardless until the fall of 1953—and say this might bear looking into," Zeidman recalled.[26] Weekly communiqués between Project Henry and Project Blue Book reveal that Hynek felt no reluctance in wheedling for greater resources to investigate those cases that appeared to have some scientific value. This seemed realistic, since First Lieutenant Robert Olsson, a relative pushover, was placed in command of Project Blue Book while Captain Ruppelt was temporarily reassigned to teach at an air force intelligence school.

"Hynek was paid to investigate only reports allocated to him by Blue Book," Zeidman explained, noting that anything else would not be reimbursed. "Many times we asked people to send a report in to Wright-Patterson so that a case we were already working on (privately) and had spent money investigating could become 'official.' Sometimes it worked, sometimes it didn't."[27]

Every week or two, Hynek would visit the Blue Book offices at

Wright-Patterson, and now and then Zeidman would tag along. In her 1991 reminiscences, Zeidman painted a picture of the Blue Book operations that could only dishearten UFO aficionados. "The Blue Book facility—building 263, not Hangar 18—consisted of three cramped, crummy little offices," she wrote. "The paint was peeling, and the file cabinets were warped. There were a United States map with pins stuck in it, a sergeant gofer, a gum-cracking, beehive-hairdoed secretary (a civilian), and a dried-out coffee pot on the window sill." Inside those warped cabinets, case files were arranged chronologically, making any other searches an exercise in frustration. "No wonder," she concluded, "I never saw Capt. Ed Ruppelt, the Blue Book head, smile."[28]

While Project Blue Book, in its dingy little offices, was busy trying to convince Americans that it was impossible for spaceships from other worlds to be visiting Earth, a growing number of science enthusiasts set about convincing Americans that human spaceflight was not only possible but right around the corner.

The movement had started in earnest in 1949 when science writer Willy Ley teamed up with illustrator and Hollywood special effects artist Chesley Bonestell to give the public a realistic conception of how man might escape Earth's gravitational field and venture forth into the solar system. Their book, *The Conquest of Space*—described by science fiction writer Arthur C. Clarke as "a delight to the eye and a stimulus to the imagination" and "something of a popular science best-seller in the United States"[29]—captivated readers with its sober depiction of mankind's first voyages to the moon, Mars, and, inevitably, the outer planets.

Ley had been writing popular science articles on the wonders of rocketry and space travel since immigrating to the United States from Germany during the rise of the Third Reich, and Bonestell had made a name for himself when a series of his paintings depicting spectacular close-ups of Jupiter and Saturn were published in *Life* magazine in 1944. Together, they made a formidable literary pair. "The realism of [Bonestell's] paintings coupled with Willy Ley's confident expository prose, convinced a generation of post–

World War II readers that space flight was possible in their life-time."[30]

On the strength of the *Life* piece, *Collier's* magazine drafted Ley and Bonestell—along with America's new rocketry chief Dr. Wernher von Braun and a team of his fellow Peenemünders—to create a series of illustrated articles promoting manned spaceflight.

"Man Will Conquer Space Soon," announced the cover head-line on the March 22, 1952, issue, and lest readers think that pre-diction too bold or fanciful, it was bolstered by the assurance that "Top Scientists Tell How in 15 Startling Pages." The cover image was a stunner: a full bleed of a Bonestell illustration depicting a winged precursor to the NASA Space Shuttle shedding its first stage in Earth orbit, readying to conquer anything space might throw at it. The series proved so popular that *Collier's* kept pub-lishing sequels and variations for another two years, and even-tually expanded the pieces to book-length. In between *Collier's* commissions, Bonestell also provided photorealistic depictions of the moon's surface for the 1950 science fiction movie *Destination Moon* and depictions of Mars for its semi-sequel, *Conquest of Space,* itself directly inspired by Bonestell and Ley's 1949 bestseller.

Not to be left behind, animator and movie mogul Walt Disney hired Ley and von Braun to create a series of television films based on their previous works. Disney's three-part animated series, be-ginning with *Man in Space* in March 1955, was a curious offering from the man who brought the world *Snow White and the Seven Dwarfs* and *Fantasia,* but Disney was a bit of a futurist himself; in fact, his trio of TV specials, over and above their sheer educational and entertainment value, were extended-length commercials for the Tomorrowland attraction at his new theme park in Anaheim, California. Tomorrowland, dominated by the gleaming white *TWA Moonliner* rocketship seen in *Man in Space*—so tall it tow-ered over even the highest spire on Cinderella's Castle—offered Disneyland visitors a trip into the streamlined future of "1986." Once there, the curious time travelers could take a simulated voy-age into space and see Earth from high orbit.

Ley, writing in the October 1955 issue of the popular science fiction magazine *Galaxy,* said of Disney's efforts, "This is how a very large portion of the public of the United States and Canada got to see a first lesson in the principles of space travel, or rather of the beginnings of space travel, the first steps outside the atmosphere."[31] So influential were Ley, von Braun, Bonestell, and Disney, that "by the mid-1950s the subject of spaceflight permeated American culture."[32]

CHAPTER 7

HYNEK IN WONDERLAND

THE YEAR 1953 MARKED A DRAMATIC TRANSITION IN THE COLD WAR, the existential political confrontation that had defined relations between the United States and the Soviet Union since the end of World War II. That year, Americans saw newly elected Dwight D. Eisenhower sworn in as their president, while Joseph Stalin, whose twenty-five-year domination of the USSR was marked by bloody purges of his political enemies, passed away.

The situation resulted in a subtle yet significant pullback on both sides, as the new leader of the Communist Party of the Soviet Union, Nikita Khrushchev, consolidated his power in the halls of the Kremlin, and Eisenhower, the first Republican president in nearly two decades, sought to reduce America's military spending and adopted a strategy of "brinksmanship," in which the mere threat of massive military engagement was enough to deter both sides from pushing the button.

This new strategy did little, if anything, to calm the fears of Americans, who were still anxious and confused by the fact that the Allies' tremendous victory in World War II had not guaranteed them peace and security in their homes, their communities, and their "beautiful for spacious skies." It was the skies that worried Americans: men, women, children, and members of the air force alike.

"I HAVE THOUGHT ALL THE KNOWN QUANTITIES over in the case of these objects and arrive at a not too unreasonable answer. The answer is that these objects are Unknown, definitely and positively . . . the objects sighted actually existed, that they were not aircraft (as we know them), not stars and not balloons. However, with my limited knowledge of the subject I could not say what they are."[1]

This admission of abject failure was written by a nervous American whose one and only job duty was to watch the skies and identify unknown aircraft. Staff Sergeant Wesley N. Harry was in charge of the U.S. Air Force Ground Observer Squadron in Bismarck, North Dakota, when he admitted to his puzzlement in a letter to Hynek. "One thing does stand out in my mind though and that is, that little is known of the Universe and it would be possible for another world to make technical advancements much farther than our own civilization."[2] The unknown objects that inspired Sergeant Harry's extraterrestrial musings and gave Hynek his first experience as a UFO field investigator might have gone unnoticed were it not for the climate of the Cold War and the strategy of brinksmanship. In 1953, the United States' skies were not yet guarded by radar defenses, and the task of watching for approaching Soviet bombers fell to a dedicated army of civic-minded sky watchers known as the Ground Observer Corps (GOC). "They were teenagers and housewives, manning search towers and bare rooftops, equipped only with binoculars," explained one military reporter. "Through the war years and most of the 1950s, GOC members spotted and plotted the movements of potentially hostile aircraft."[3]

Phyllis Killian and her binoculars kept watch over the small town of Blackhawk (sometimes spelled Black Hawk), South Dakota, just seven miles northwest of Rapid City. Killian was an experienced aerial observer and was regarded as one of the best reporting in to the Rapid City GOC Filter Center. "She is reputed to know planes in detail," Hynek was to observe, "and has on occasion identified the aircraft as to type and motor when it would only dimmly [*sic*] be heard in the distance."

On the night of August 5, a few hours into her shift, Killian's attention was drawn to a red light low on the horizon in the southeastern sky. She was at that moment visiting with her neighbor, a Mrs. Daughenbach, who also saw the red object and thought it was the light atop a nearby radio tower. This notion was quickly laid to rest when the light changed from red to green and started to move to the south-southwest.

Killian dutifully reported the sighting to the Filter Center in Rapid City, and from there an alert was sent to the radar operator at Ellsworth Air Force Base, twelve miles to the east. According to Captain Ruppelt, "There was a target exactly where the lady reported the light to be."[4]

As Warrant Officer Junior Grade Howell I. Bennett tracked the object on radar, he reached Killian on the phone and then asked three airmen to step outside the control tower and look for the object, which they quickly sighted. At the exact moment that Killian reported to Bennett that the object was now moving toward Rapid City, the three men outside the tower reported the same. "The controller looked down at his scope—the target was moving toward Rapid City," Ruppelt reported.[5]

"The object was round, changing colors from bright red to Greenish Silver depending on speed," Bennett later explained. "There was no sound that could be heard although they stated it looked like a 'V' shaped vapor trail was following the object. The maneuverability of the object is very sharp in all directions, such as up and down and sideways."[6]

"A master sergeant who had seen and heard the happenings told me that in all his years of duty—combat radar operations in both Europe and Korea—he'd never been so completely awed by anything," Ruppelt wrote. "When the warrant officer had yelled down at him and asked him what he thought they should do, he'd just stood there. 'After all,' he told me, 'what in hell could we do—they're bigger than all of us.'"[7]

An F-84 was already airborne at the time of the sighting, and

Bennett quickly vectored the plane to intercept the object, which was now at sixteen thousand feet. When Lieutenant John W. Stockman closed in on the object, which he described as "bright silver in color," it moved off at a high rate of speed.

Stockman pursued the object relentlessly but was never able to get closer than three miles from his target. Every time the pilot approached the object, it skipped away, and when Stockman found himself 120 miles from base, he realized he would have to turn back for refueling. "When I talked to him, he said he was damn glad that he was running out of fuel," Ruppelt wrote, "because being out over some mighty desolate country alone with a UFO can cause some worry."[8]

Stockman could not have been reassured, however, when the object suddenly reversed course and started to follow him back to the base.

Another F-84 was ready to scramble from Ellsworth, so Bennett sent the second pilot, Lieutenant David K. Needham, off after the object, now eighty-five miles away. Needham gave chase, but once again the light—now alternating between white and green—kept its distance with apparent ease. Nonetheless, Needham did get close enough for his gunsight radar to lock on the light, indicating that it was a solid object.

"Then he was scared," Ruppelt wrote.[9] Needham had faced off against some of the deadliest fighters in the world over Germany and Korea, but this elusive, amorphous craft spooked him like no other. Putting his macho image at risk, Needham asked the controller if he could break off the intercept.

By this time, several other lights had been spotted in the sky above Rapid City, but only the primary object pursued by the two pilots shot off to the northeast after Needham abandoned his pursuit. With nothing else to do, Bennett made sure that "all GOC Filter Centers were alerted in the northern area"[10] that an unidentified object was headed their way.

At the Bismarck GOC Filter Center, more than two hundred miles away in North Dakota, Sergeant Harry stepped out onto the roof of his building and spotted Bennett's object approaching from the southwest. "It definitely stood out in the night sky as seeming to be at least twice the size of any of the thousands of stars out at the time, and because its color was changing," he reported.[11]

Several other observers joined Harry on the roof of the Filter Center as the object hovered in the western sky and appeared to bob up and down. "By this time everybody at the Filter Center was very excited and, as they said, shaking," Hynek wrote later. "They had the feeling that somebody was watching them."

Between midnight and one A.M. the object had disappeared over the horizon to the west-northwest, only to be replaced by a trio of similar floating lights to the east. By now Jack Wilhelm, the radar operator at the local civilian airport, had picked up the objects as well, and, just as in Rapid City, visual sightings were instantaneously verified by radar returns.

Sergeant Harry firmly believed the objects were not conventional aircraft, because they made no noise and displayed no aerodynamic features, and there was no record of any known aircraft in the sky at that time. A thorough check had confirmed that no weather or test balloons had been launched in the vicinity that night, eliminating that possibility as well. Furthermore, Harry believed that the objects' erratic movements made it unlikely that they were stars.

Eventually a "known" aircraft did appear on the scene, which added to the mystery. As clouds moved in over Bismarck, a massive U.S. Air Force C-124 Globemaster heavy-lift transport plane lumbered through the sky on its way to parts unknown and seemed to elicit an almost *conscious* response from the unidentified flying objects.

"At this time, prior to even knowing of the C-124 in the area, I noticed the pulsating signal which would last about 5 seconds, stop for approximately 15–30 seconds and begin again," Harry said of what appeared to be a message passing between the UFOs. "This happened several times and was witnessed independently by 5 other people from vantage points in and around the Filter Center. These signals or whatever they were lasted only until the C-124 was out of the area."[12] The crew members of the C-124 were alerted to the presence of the objects, but they saw nothing. The gathering blanket of low clouds was blamed for the lack of visual contact.

When the reports came in to Project Blue Book from the Bismarck Filter Center and Ellsworth Air Force Base, Captain Ruppelt was about to resume life as a civilian. Technical intelligence officers were no longer "on the critical list" at the air force, and Ruppelt's talents were no longer needed. "The UFO's must have known that I was leaving," he quipped in his book, for the sighting reports came in the very day he learned of his discharge.[13] With such a tantalizing case file on his desk and no successor yet named, Ruppelt caught the next flight to Ellsworth AFB and conducted his final field investigation.

It immediately becomes clear upon reading his report that the words "excited" and "excitable" were Project Blue Book shorthand for "unreliable." By the time the Blue Book investigation was complete, virtually every key witness at both the South Dakota and North Dakota locations had been tarred with this epithet at least once—and often by each other.

Case in point: after conferring with the base operations officer at Ellsworth, Ruppelt met with Warrant Officer Bennett, Lieutenant Stockman, and Lieutenant Needham, but not with Phyllis Killian, the ground observer who first reported the UFO. "No attempt was made to contact the GOC observers at Blackhawk," Ruppelt noted in his report. "They had been interrogated by base personnel and were 'all excited.' It was believed Capt. Ruppelt's talking to them would only further excite them needlessly."

Ruppelt was unable to make much sense of the South Dakota incident upon his arrival at Ellsworth. He felt that the two pilots were unsure of what they had seen, while Bennett, the radar operator, was "trying to prove the existence of a UFO." He was no more impressed with the airmen who were sent outside by Bennett to observe the object, stating that "they were told to go out and look for a light so they saw one."

Clearly at a loss to explain an occurrence that he would later describe as "nothing but a big question mark,"[14] the captain closed his report with this bombshell: "It is recommended that further study of this incident be made by Dr. Hynek and A/lC Futch." Never before had a Blue Book project chief suggested that a UFO event was worthy of further study, or suggested the involvement of the astronomical consultant in the investigation of a case.

PERHAPS RUPPELT WAS FEELING the heat from the competition. For several months, two new civilian UFO research groups had been nipping at Blue Book's heels, and the air force was not used to being forced to guard its turf.

The Aerial Phenomenon Research Organization (APRO) was formed in 1952 in Sturgeon Bay, Wisconsin, by flying saucer enthusiasts Jim and Coral Lorenzen. In California, meanwhile, aviation writer Ed Sullivan founded Civilian Saucer Investigation of Los Angeles (CSI). Both organizations quickly became irritants to Ruppelt and Blue Book: here were these upstart, underfunded, poorly organized clots of untrained amateur UFO investigators, openly challenging the air force to do a better job of studying the phenomenon, and giving voice to a growing sentiment among the American public that the authorities couldn't be trusted to tell the truth about the UFO phenomenon.

While CSI would burn out in only two years, APRO proved to be a formidable force, boasting three thousand members at its peak and remaining an active player in the UFO field for three decades. Technical writer and editor Coral Lorenzen had seen two UFOs in her lifetime; not only could she speak with authority on the subject, she genuinely understood the impact a UFO experience could have on a witness's life. She and her husband and partner, Jim, an electrical engineer, openly embraced the extraterrestrial hypothesis and actively investigated "UFO occupant" cases, a rare variety of "excessively exotic"[15] UFO event that other researchers—Hynek included—handled with kid gloves, if at all.

At the same time that Project Blue Book and the civilian UFO groups were facing off against each other, both found themselves having to deal with the annoyance of the "Contactee" phenomenon, which simply appeared without warning in 1953 and took the nation by storm.

Contactees were ordinary, undistinguished men and women who claimed to have been chosen for intense, highly personal, on-going interactions with aliens from distant planets. The key distinction was that Contactees claimed to have been *invited* aboard the flying saucers, so they did not see fit to report their experiences to the air force or the civilian UFO groups; instead they went straight to the public, writing books, giving lectures, and granting interviews to the press. In so doing, the Contactees passed on the wisdom of the aliens to the human race, as they had been instructed to do.

Contactee George Adamski, who ran a commune and a burger stand near Palomar Observatory in Southern California, claimed to be on close terms with the angelic, vaguely Scandinavian alien Orthon, who warned Adamski that the human race must give up its violent addiction to war. Aeronautical engineer George Van Tassel's alien friend, Ashtar from the Council of Seven Lights, meanwhile, told him to host a UFO convention at his home in Yucca Valley, California, and then to build the "Integratron," a dome-shaped structure that could, the alien claimed, rejuvenate human cells (the structure still stands today—Van Tassel was not so lucky—and remains a popular spiritual retreat). Mechanic Truman Bethurum was ushered into a flying saucer and taken on a flight through space by the friendly humanoids of the planet Clarion. The captain of the ship, the beautiful Clarion woman Aura, repeated Orthon's message of world disarmament.

The serious UFO groups could not hope to compete against so much razzle-dazzle. "In media coverage, colorful contactees quickly trumped serious UFO investigators—who term them-selves 'ufologists,'" wrote author Fred Nadis. "Long John Nebel's all-night talk show on WOR in New York, dedicated to discus-sions of voodoo, beatniks, perpetual motion, the paranormal, and the otherwise bizarre, often featured conversations with Adamski, Van Tassel, and Bethurum." Even Steve Allen's *The Tonight Show* gave the Contactees a friendly venue from which to spread their messages from the stars.[16]

Given the opportunity, Hynek could not resist exposing a Contactee as, if not a fraud, then a fool. Once, at the height of Adamski's popularity, Hynek found himself in a car with a group of fellow astronomers driving to the observatory on Mount Palomar, and on a whim they stopped at Adamski's diner for a bit of fun. As they waited for their burgers, the astronomers spoke very loudly and excitedly about some new optical filters for the two-hundred-inch Hale Telescope at Palomar that would yield indisputable proof that there was life on Mars.

Sure enough, the Contactee bit.

"Oh! were Adamski's ears flapping in the breeze!" Hynek said. "He came over and introduced himself, and I'm sure he told people afterwards that he had it straight from the top astronomers that there was life on Mars!"[17]

When Hynek engaged Adamski in conversation, however, he was in for a letdown. The king of the Contactees didn't want to talk about astronomy or Mars at all, Hynek said: "All he wanted to do was to sell me some pictures!"[18]

Despite the pressure from the civilian competition, Ruppelt's recommendation for further investigation of his final case had to be temporarily shelved. That summer Hynek was preparing for a demanding six-week European trip, and his golden opportunity with the Blackhawk and Bismarck UFOs would have to wait.

STRANGE PHENOMENA IN THE SKY seemed to have a transformative effect on Hynek's destiny, and in 1953 the heavens were once again plotting to reshape his life. This time the culprit was a total eclipse of the sun that was to occur on June 30 of the following year. The totality of the eclipse was to sweep from the American Midwest up across Labrador, Greenland, and Iceland, then across Scandinavia, and finally taper off across Iran and Central Asia. The astronomy department of The Ohio State University, at the behest of the U.S. Air Force's Cambridge Research Center, was planning to set up four observation stations along the path of the totality, and Hynek, as scientific liaison, was responsible for diplomatically paving the way.

Because the expeditions were underwritten by the air force, Hynek needed to assure his European colleagues that the monitoring stations were being established for purely scientific—not military—reasons. "In a nutshell, I went over to establish scientific co-operation with astronomers in Sweden and Finland," Hynek said. "It's a move in international co-operation in science,"[19] a cause that had been near and dear to his heart since the war.

As Hynek explained it, OSU astronomers were planning to take very accurate measurements of the moon during the eclipse and attain a better measure of Earth's size and shape as well. In addition to explaining these objectives to scientists in other countries, he was responsible for the calibration and accuracy of the equipment to be used at the four observation posts. "Split second timing will be achieved between American observation points and those in Scandinavia by the use of crystal clocks calibrated by radio. Each frame of photography will be timed down to 1/100 of a second."[20]

After visiting with colleagues in Sweden and Finland, Hynek still found time to vacation for a few days in England, France, and Denmark. Winding down after several weeks of intense technical work, it came as a true pleasure for Hynek to take a step back in time while he was in France. While there, he had the pleasure of peering through Napoleon's portable telescope—one of the high points of the trip.

In addition to laying the groundwork for the OSU eclipse expeditions, Hynek was busy penning a weekly astronomy column for the *Columbus Dispatch*. Dubbed "Scanning the Skies," Hynek's column alerted local stargazers to the upcoming week's celestial displays, but it often took side trips into whatever scientific curiosity caught its author's fancy. One week the column might deal with the age of the universe, and the next it might explore the nature of comets.

In one edition of Scanning the Skies, the doctor spent some time musing on the long absent Halley's Comet. "We honor the name of Halley because he was the first to prove that comets were not frightening supernatural phenomena but instead orderly members of our solar system," he wrote, unconsciously drawing an eerie parallel to his own UFO research. "He correctly predicted that the comet which since bears his name, would return in 1759, and showed that this was the same comet that had been here many times before, as for example, in 1066, at the time of the Norman Conquest."[21]

Noting that the comet had recently begun its return journey from the depths of space, Hynek predicted that "those of us who are around in 1986 can expect to see a fine sight."[22]

At times Hynek was asked to elucidate the subject of flying saucers, but he was still very cautious about how he presented his thoughts about UFOs to the public. When asked about flying saucers after giving an astronomy lecture in Lima, Ohio, in 1953, Hynek lamented that "that awful, awful word" had come up. He repeated his definition of a flying saucer to his amused audience as "any observation that remains unexplained long enough for somebody to write a report about it," and he cautioned that while there was "nothing scientifically impossible" about the idea that we were being visited by intelligent beings from another world, it was "highly, highly, highly improbable."[23]

Ever since the first mass-market UFO books and alien invasion movies had taken the country by storm, Hynek had been forced to spend more and more of his time and energy fighting off wild speculation about menacing aliens from other planets. It scarcely mattered that the centerpiece of Frank Scully's book had long since been exposed as a hoax by *True* magazine; by 1953 the "saucer crash" phenomenon had taken on a spectacular, surreal life of its own, and Hynek could not escape the silliness. "The fact that [*Behind the Flying Saucers*] was a loudly bad book was beside the point," wrote the *True* reporter, opining that "it affected, in some degree, one way or another, the thinking of millions of people."[24]

Even *Life* magazine, the mighty news and cultural affairs weekly that dominated newsstands and watercooler conversations in the 1950s, was not immune to the UFO spell. When *Life* spoke, the whole country listened. And when *Life* spoke about flying saucers, as it did in the article "Have We Visitors from Space?" in its

April 7, 1952, issue, then flying saucers really meant something. "For four years the U.S. public has wondered, worried or smirked over the strange and insistent tales of eerie objects streaking across American skies," the article began. "Generally the tales have provoked only chills or titters, only rarely, reflection or analysis."[25] That was about to change, if *Life* had anything to say about it.

The story had the stupendous good fortune to be published in an issue that featured a three-quarter-bleed cover photo of the beckoning face and bare shoulders of Marilyn Monroe. For readers who ventured beyond the puff piece on Monroe, the magazine served up a provocative analysis of the UFO phenomenon and a blistering critique of the air force's efforts to deal with flying saucers.

"The project set up to investigate the saucers . . . seemed to have been fashioned more as a sedative to public controversy than as a serious inquiry into the facts," *Life*'s authors asserted, claiming that the air force's efforts should be defined not by the vast majority of UFO cases it was able to explain away, but by the thirty-four reports that left it flummoxed. "These stubborn 34, seemingly unexplainable, were briskly dismissed as psychological aberrations."[26] The authors considered ten of these unexplained cases and came away declaring that these incidents defy simple explanation and provoke "the most incredible questions."[27]

"Have We Visitors from Space?" was an explosive piece of reporting and quickly led to a dramatic uptick in UFO sightings around the country. And the air force had only itself to blame, as one of its own, an unnamed intelligence officer, contributed the article's money quote: "The higher you go in the Air Force, the more seriously they take the flying saucers."[28]

Small wonder, then, that Hollywood was following the lead of *True* and *Life,* as filmmakers targeted a suddenly booming market with increasing success. "The proliferation of science fiction films is one of the most interesting developments in post–World War II film history," wrote film historian Patrick Lucanio. "An estimated 500 film features and shorts made between 1948 and 1962 can be indexed under the broad heading of science fiction. One might argue convincingly that never in the history of motion pictures has any other genre developed and multiplied so rapidly in so brief a period."[29]

Or, as Douglas Trumbull, the special effects designer who created the UFOs in the 1977 movie *Close Encounters of the Third Kind,* put it, "No one's ever lost a lot of money making a movie about alien attacks or stories about UFOs."

The 1953 film version of H. G. Wells's *The War of the Worlds* treated its viewers to what cultural critic Susan Sontag referred to as "the core of a good science fiction film": e.g., "the aesthetics of destruction, with the peculiar beauties to be found in wreaking havoc, making a mess."[30] Here, the film's mystical scientist is described by one character as "the top man in astro-nuclear physics," if there ever was such a thing. He and the rest of the cast are summarily outsmarted and overpowered by the rampaging Martians in their tripod war machines, and they spend much of the film running for their lives. When even our H-bomb can't break through the "electromagnetic covering" shielding the Martian ships, and it looks as though their heat ray will disintegrate every last one of us, the scientist succumbs to despair . . . But, just as they did in Wells's novel, the Martians in the film fall victim to an earthly microorganism and perish before they can deal the final blow to mankind.

"These films reflect world-wide anxieties, and they serve to allay them," wrote Sontag, describing an American society that was increasingly nervous about the escalating Cold War and about the mismanagement of science by either side. To Sontag, the alien invasion movies of the 1950s marked a historically significant development in human consciousness: "From now on to the end of human history," she wrote, "every person would spend his individual life not only under the threat of individual death, which is certain, but of something almost unsupportable psychologically— collective incineration and extinction which could come any time, virtually without warning."[31]

In 1953, that collective extinction was just as likely to come from outer space as from a Soviet missile. UFOs had become a part of our history and our psyches, and soulless aliens were quickly becoming universal symbols of primal horror. "These alien invaders practice a crime which is worse than murder," Sontag wrote. "They do not simply kill the person. They obliterate him."[32]

And yet amid the destruction there could exist a slender thread of hope. "The humbling of humanity is not to be taken as a defeat but as a victory," wrote Lucanio. "The closing images of the film reveal triumph mixed with loss; the invader has been repelled but at the same time man's perception of himself and his world has been altered."[33]

UFO books, articles, and movies of the early 1950s, then, operated in concert to deliver an unsettling message: the alien invasion movies did their best to convince the public that aliens from other worlds were ready to attack and destroy us at any time, while UFO books and magazine articles did their best to convince the public that the government already knew all about the aliens' nefarious plans and was keeping them a secret. This potent dynamic reached its inevitable zenith when the film rights to Donald Keyhoe's second government cover-up book, *Flying Saucers from Outer Space,* were purchased by Columbia Pictures and used as the "inspiration" for the 1956 alien invasion movie *Earth vs. the Flying Saucers.* While there is no outright government cover-up in the film, the aliens seem to share Keyhoe's anger at politicians and authority figures. First, they murder a four-star general in a particularly grisly fashion— draining his brain of all knowledge and tossing him from the open hatch of a saucer into a blazing forest fire—then they quickly target Washington, D.C., for complete incineration.*

---

* Keyhoe was, reportedly, not pleased with the film, but if he were alive today he might appreciate its unique status as a footnote in both film and UFO history. Not only was famed special effects animator Ray Harryhausen said to have consulted with Contactee George Adamski on the design of the film's spinning flying saucers, but those very same saucers reappeared as an affectionate tribute in Tim Burton's 1996 alien invasion comedy, *Mars Attacks!* In addition, Harryhausen's scenes from *Earth vs. the Flying Saucers,* in which the saucers smash into the Washington Monument and the U.S. Capitol Building, were

It was against this backdrop of alien invasion angst that Hynek at last had his opportunity to personally investigate the Black-hawk and Bismarck sightings from the previous August.

When Hynek arrived in Bismarck, he interviewed Sergeant Harry, civilian radar operator Jack Wilhelm, and the other ground observers who had joined Harry on the roof of the Filter Center that night. He was impressed with Harry's description of the positions and movement of the objects, in part because the sergeant had taken care to position himself where he had nearby power lines serving as a spatial reference point, and because he had thought to carve outlines of his shoes to mark where he stood on the tar roof of the building. The other witnesses impressed him as well, especially Wilhelm: "Wilhelm has had much opportunity to observe stars from the control tower, and it appears inconceivable to me that a man with 13 years [of] experience could have mistaken stars for these objects."

Hynek was particularly struck by Harry's description of the "pulsating signal" that had been observed passing between the UFOs as the C-124 flew by: "It was likened to the signaling of a ship's signal blinker but not as fast."

Hynek's confident analysis of the Blackhawk–Bismarck case marked a major turning point in his Blue Book career. When given the opportunity to investigate a case on his own, Hynek felt free for the first time to openly question, even criticize, his former boss's investigation. One of Ruppelt's decisions in particular rankled Hynek. Despite his positive feelings about the South Dakota witnesses, Ruppelt had reasoned that it would not be worth four days of extra travel to interview the witnesses in North Dakota. "All the sightings at Bismarck are doubtful," he had reported. "The AC&W Station called the Bismarck Filter Center and told them to 'look for flying saucers,' a perfect set up to see every star move around."

---

referenced in another 1996 alien invasion film, *Independence Day,* in a scene—the film's trademark image—in which the White House is obliterated by a flying saucer.

Hynek was incensed. "I cannot accept this interpretation in the least since the [Bismarck] observers interrogated were in my opinion a definite cut above the average person," he reported. "The operation of the Filter Center seemed to be very efficient and it was apparent that they had recruited very good people for the work. Sgt. Harry in particular appears thoroughly reliable and accurate and is very much 'on the ball.'"

He felt very strongly that something real had occurred at Bismarck. "We can conclude that there is no ready explanation for these objects. To call them aircraft, balloons, or stars becomes somewhat of a desperation move, though the balloon hypothesis would be the least objectionable," he wrote, ultimately concluding, "Until further explanation is forthcoming, this incident must be put down as one of the best examples of the 'nocturnal meandering lights' which have been reported elsewhere in the literature. To do otherwise would be to doubt the first order objectivity of observers who stood up very well under considerable questioning."

His investigation in North Dakota completed, Hynek went on to Rapid City to investigate the South Dakota sightings, and once again he found himself in utter disagreement with his boss. Where Ruppelt had found the South Dakota witnesses generally credible, Hynek found the whole lot woefully disappointing.

He did not warm up to Phyllis Killian, the ground observer, describing her as "an excitable and rather cocky individual, and in some respects overbearing."

Hynek was also exasperated by an "interesting personality conflict" that emerged between Ellsworth AFB radar operator Bennett and pilot Needham. Essentially, both men accused the other of being "excitable." Needham further alienated Hynek with his unwillingness to be pinned down on the details of his story: "He did not wish to commit himself on anything, and gave the impression that the less he heard of it in the future the better."

The unreliability of the South Dakota witnesses was underscored by a report from Major Allen L. Atwell of Ellsworth AFB, sent to Hynek a few days after his visit. Atwell had conducted a thorough investigation of the personnel involved in the UFO sighting and found little reason to believe any of them: "Considering the reliability of the military witnesses at the radar site, the excitement of Mr. Bennett, and the relatively poor description of subsequent sightings in the Rapid City area, I feel that the incident can be discounted from consideration entirely," Atwell wrote.[34] Hynek was in full agreement. "We are ready to write off the Rapid City–Black Hawk incident as 'information from questionable sources,'" he wrote back to Atwell. "In short, we can't trust the evidence, and when this happens in any scientific procedure the thing to do is throw the whole thing out the window . . . I think we shall probably proceed by discounting the Rapid City Incident altogether and concentrate on the Bismarck incident."[35]

Despite his certainty, Hynek was presented with a thorny problem: If the South Dakota sighting was so flimsy and the North Dakota sighting was so credible, how was it that the former appeared to lead directly to the latter? "As far as the Rapid City sightings are concerned, we have a train of circumstances which, if entirely coincidental, are remarkable indeed," Hynek stated in his contorted Blue Book report.

"The entire incident, in the opinion of this observer," wrote a frustrated Hynek, "has too much of an Alice in Wonderland flavor for comfort."

Hynek's first field investigation, then, was an "unknown," and a frustrating one at that. "Should not the question be asked, 'Why, after three hours of nearly continuous sightings, was the Air Defense Command unable to identify, intercept, or otherwise cope with the event?'" he wrote in a lengthy addendum to his Blue Book report. He continued: "Is it too much to ask that a defense post which is alerted and has access to unknown objects for about three hours, should in that period of time be able to tell at least whether the objects were tangible?"

By the time Hynek filed his report, Donald Keyhoe's second book, *Flying Saucers from Outer Space,* had come out and Hynek was well aware of the danger it represented to Project Blue Book. "The consensus among the public is, I would say, that the Air Force 'knows all about it' but refuses to say anything," Hynek told his superiors. He couldn't help pointing out that in cases such as the spectacular sightings over Washington, D.C., the previous year, Keyhoe had actually done better, more thorough research than had the "official" air force investigators, "a dangerous precedent indeed."

As Keyhoe's book racked up higher and higher sales numbers, Hynek's life was all but taken over by the solar eclipse project at Ohio State. As summer approached, mission creep set in and the plans expanded dramatically. No longer would the expeditions be going to Sweden and Finland; to take full advantage of the last full solar eclipse to cross both North America and Europe until 2151, now OSU would send four full scientific teams to Knob Lake, Quebec; Labrador; Greenland; and Bandar Shah, Iran. Placed in charge of the Iranian expedition was one Dr. J. Allen Hynek, director of McMillin Observatory.

"Shortly after dawn on June 30, four teams of Ohio State scientists will concentrate on the rising sun. Three and a half minutes later, the field work will be done," reported the *Ohio State University Monthly*. "After that, there will be nothing to be done but re-pack the equipment and return to the University. But the results of those three and a half minutes of work will tell the Air Force trans-oceanic distances in more accurate terms than hitherto known."[36]

The article went on to explain that with photographs of the eclipse taken at precise times at the four outposts, OSU scientists would be able to work out the distance between Europe and the United States "to an accuracy within 100–150 feet."[37] Previous measurements came no closer than 500 to 600 feet.

Because each team would need close to a month at its selected observation site to prepare and calibrate its equipment, Hynek had a lot to write about in his newspaper column. As early as January 1954, he was talking up the preparations for the mission in Scanning the Skies and describing what the expeditions would witness: "For a few moments night will fall in broad daylight and the stars will come out. Chickens and birds will start to go to roost, only to be rudely disturbed by a new day a few minutes thereafter! And above all, the sight that very few people ever see, the pearly white corona of the sun will be visible for a little more than two minutes."[38]

"If you can make a quick auto trip to the 60-mile-wide path the moon's shadow will make across northern Wisconsin and Michigan, do it by all means. It will be a chance of a lifetime," he exhorted his readers in his June 6 column. "Mrs. Hynek, for instance, is taking all three children in the car to see the eclipse. Weather permitting, they'll see it 2½ hours before I do, in Iran."[39]

Hynek's arrival in Iran was a jarring experience. He found the village of Kafshgiri, nestled at the southern tip of the Caspian Sea, to be a pleasant place geographically, "but there the description ends."[40]

"We did not find the water we had been informed would be available," he wrote on June 15. "Brushing one's teeth with soda pop (we managed to pick up a few bottles in Gurgan) was a new experience. An underground well has now been located, and with copious quantities of chlorine tablets, death by thirst or dysentery has apparently been avoided."[41] Adding to the research team's worries were "extremely curious" townspeople, who had to be kept out of the eclipse camp by Iranian army guards, and rumors that the eclipse would bring about massive destruction in the form of earthquakes and tidal waves.

On the lighter side, the railroad trip through the mountains from Tehran to Bandar Shah turned out to be a real-life thrill ride for Hynek and his team. After the train reached the pass at an altitude of eight thousand feet, the locals all leaned out the windows in anticipation . . . "I wondered what was up. Plenty!" Hynek burst out. "The engineer starts on a joy ride down the other side toward Bandar Shah. It has all the aspects of a train ride in an amusement park. In about 30 minutes the train—at a fair clip—goes through no less than 40 tunnels and around 20 curves, as it winds continuously downward on a 3 per cent grade."

In the next Iran update from Hynek, he revealed to his readers that his eclipse camp had been built on the top of a fortress that dated back to the time of Alexander the Great, more than two thousand years ago. Working conditions were unpleasant, to say the least. The Iranian drivers terrified the scientists with their disregard for speed limits, pedestrians, other vehicles, and roads. The heat was so oppressive that the team members rose at 5:30 so that they could complete the day's work by 10 A.M. Anything left to be done after that had to be tended to in the late afternoon.

Hynek's final report from Iran is crushing: "To scanning the skies readers: It's a sad day in Kafshkiri [*sic*] today. The great eclipse has struck out."[42]

After the months of planning, after all the travel and calculating and diplomacy and obsessing, the entire mission was done in by the age-old enemy of the astronomer: "A black cloud, pitched from out over the Caspian Sea, at just the critical moment covered the sun and hid from view the sun's corona, which we had traveled all the way to Iran to see," Hynek lamented. "A clear case of celestial sabotage!"[43]

What's worse, the three other observation posts along the path of the eclipse were all clouded over just as the moon passed before the sun. Hynek's dismay was amplified by the knowledge that the next eclipse to provide a similar opportunity was 197 years distant. The failure of the mission was absolute. This time no flashlight could outwit the weather. "When the clouds came no one said a thing," he wrote. "There was no need for words. Our disappointment was too intense for words."[44]

What good was Hynek's faith in the laws of physics and astronomy when those very laws could so easily obscure his vision? It was a conundrum that would drive him for the rest of his life.

CHAPTER 8

FLYING SAUCER CONSPIRACY

SINCE HIS COLLEGE DAYS, and perhaps before that, Hynek had held a special fascination with Johannes Kepler (1571–1630), the German astronomer, mathematician, and astrologer who conceived of the Three Laws of Planetary Motion and, thus, deciphered the workings of the solar system.

Two childhood memories recorded by Kepler in the family horoscope had a familiar ring to Hynek: at age six, Kepler noted, "I heard much of the comet of that year, 1577, and was taken by my mother to a high place to look at it." Then, at age nine, "I was called outdoors by my parents especially, to look at the eclipse of the moon. It appeared quite red."[1]

As an adult Kepler strove to find harmony in the heavens, searching for the reason the six planets (only Mercury, Venus, Earth, Mars, Jupiter, and Saturn were known in his time) moved about so freely in the night sky while the stars remained fixed. How, he wondered, could the motion of the one be reconciled with the uniformity of the other?

These notions of movement and immobility in God's creation were of immense scientific and theological weight in the seventeenth century. The fixed stars were clearly held in place by some sort of framework, but then what power swept the planets through the sky? In their seminal book on archaeoastronomy, *Hamlet's Mill,* historians Giorgio de Santillana and Hertha von Dechend described how thinkers of Kepler's time perceived the stars and planets: "The fixed stars are the essence of Being, their assembly stands for the hidden counsels and the unspoken laws that rule the Whole. The planets, seen as gods, represent the Forces and the Will: all the forces there are, each of them seen as one aspect of heavenly power, each of them one aspect of the ruthless necessity and precision expressed by heaven. One might also say that while the fixed stars represent the kingly power, silent and unmoving, the planets are the executive power.

"Are they in total harmony?" the historians asked. "This is the dream that the contemplative mind has expressed again and again, that Kepler tried to fix . . ."[2]

AFTER SEVERAL AMBITIOUS ATTEMPTS to make sense of the motion of the planets had failed, Kepler found himself in desperate need of intellectual support. "There was only one man alive in the world who possessed the exact data which Kepler needed: Tycho de Brahe," wrote Kepler biographer Arthur Koestler. "All his hopes became now focused on Tycho, and his observatory at Uraniburg, the new wonder of the world."[3]

From his observatory, Brahe conducted nightly observations of the stars and planets, and over time he amassed a vast library of tables and measurements that plotted the exact motions of the planets over many years. It was this data that Kepler coveted, and when Brahe offered Kepler a position, his course was clear.

Brahe could make no sense of his observations of the orbit of Mars, and he put Kepler to work on the problem immediately. According to author Colin Wilson: "For the past six years [Kepler] had worked in isolation—a loner, working with inadequate data. Suddenly, he had data by the cartload. And it led to problems that were so great 'that I nearly went out of my mind.' But he saw at once that Tycho lacked the organizing intellect to make proper use of all this material, and that he, Kepler, was destined to be the architect of the grand design.

"Tycho himself realized this . . . He clung to his data like a miser, releasing it in dribs and drabs," and as a result, "the two men seemed made to detest one another."[4]

Despite Tycho's stinginess, Kepler did indeed become the architect of modern astronomy, using Tycho's thirty-five years of observations to show that: 1) the orbit of a planet is an ellipse, not a circle, and the sun is only one of its two foci; 2) a straight line between a planet and the sun always sweeps out the same area of space in a set unit of time at any point on the planet's orbit; and 3) the ratio of the square of the period of a planet to the cube of its average distance from the sun is always equal between two planets. These three laws, over time, led to Newton's discovery of gravity, and defined what Koestler described as the "watershed" in our understanding of the cosmos.

Like Kepler, Hynek had "data by the cartload," thanks to the air force's Project Blue Book files. And like Kepler, Hynek chafed at the miserable circumstance of having to serve a disagreeable master in order to gain access to that data, realizing as he did that his master "lacked the organizing intellect" to make use of all its material. But, also like Kepler, Hynek knew that in the end, the quality of the data transcended all other considerations, so he continued on with Blue Book, even when common sense might have suggested a different path, even when the inherent compromises of his position threatened to slaughter his professional reputation.

The curious parallels between the two lives don't end there. After Tycho's bizarre exit—he died of a ruptured bladder after a night of excessive drinking without the ability to avail himself of the toilet—Kepler nearly lost access to Tycho's data. When Tycho's jealous son and one of his assistants attempted to block Kepler's access to the journals, Kepler's response was to launch a preemptive strike and steal the records himself. In a similar way, Project Blue Book files had a way of somehow migrating in bulk from Wright-Patterson to Hynek's office.

Later in his career, Kepler lent his support to Galileo when the latter announced the discovery of four new planets—later discovered to be four of the moons of Jupiter—through the use of a revolutionary new device called an "optick tube," or telescope. Now, the first telescopes were rather chancy affairs; there were no mountings, the visual field was minuscule, and the optics were often imprecise, creating haloes and double images. So when other scientists looked for Galileo's new satellites, they just saw blurry smudges of light, "optical illusions in the atmosphere,"[5] or nothing at all. As a result, few believed Galileo's claims.

Kepler, however, issued a public statement in support of his colleague, even though he, Kepler, had never even seen a telescope and therefore had never seen Galileo's new "planets." The significance to Hynek of Kepler's actions is made explicit in a passage in Koestler's biography of Kepler, in which he describes the situation thus:

> The whole controversy about optical illusions, haloes, reflections from luminous clouds, and about the unreliability of testimonies, inevitably reminds one of a similar controversy three hundred years later: the flying saucers. Here, too, emotion and prejudice combined with technical difficulties against clear-cut conclusions. And here, too, it was not unreasonable for self-respecting scholars to refuse to look at the photographic "evidence" for fear of making fools of themselves . . .
>
> The Jupiter moons were no less threatening to the outlook on the world of sober scholars in 1610, than, say, extra-sensory perception was in 1950.[6]

Hynek had a unique manner of expressing his affinity to Kepler. His son Paul recalled a "wonderful father-and-son moment" from his youth when his dad told the story of Kepler and Brahe in his own wry way.

Paul had seen the name Tycho Brahe on one of his dad's test papers and thought it must be a gag. "Dad said, 'You don't know who Tycho Brahe was?' He was the astronomer for the Holy Roman emperor, and the Holy Roman emperor came one night, and he had a steak dinner and Tycho Brahe was there, and at the time Tycho Brahe was a real character, as you know—he had his nose sliced off in a duel and had a variety of fittings that he would put on it, copper or silver or whatever.

"My dad told me that Tycho Brahe was fantastic at observations, but not at math. And he had this young bright assistant who would go on to some renown with his Laws of Planetary Motion.

"So, at the dinner, Brahe, who was a drunkard, drank copious amounts of beer and had to go to the bathroom. But you don't just excuse yourself to go to the bathroom when the Holy Roman emperor is there . . . So [Dad] says that Tycho Brahe basically died because his bladder burst. He died at the table . . . and because he died, it released the data to Kepler, to formulate the Three Laws of Planetary Motion. So my father taught me that bladder bursting was great for science.

"My father was very good at turning phrases."

EIGHT YEARS AFTER THE KENNETH ARNOLD SIGHTING, with no tangible proof yet produced of the physical reality of UFOs, the researchers at Battelle Memorial Institute quietly wrapped up Project Stork and released their findings to the public in what was referred to as "Special Report #14." The October 25, 1955, press release announcing the close of the study was issued from Secretary of the Air Force Donald A. Quarles, stating boldly that "no evidence of the existence of the popularly-termed 'flying saucers' was found."[7]

Project Blue Book was off the hook.

Hynek welcomed the news with a mixture of relief, resignation, and perhaps just a little exhaustion. In a letter of congratulations to Captain Charles Hardin, his Blue Book project chief, Hynek wrote, "I personally do not feel that my time spent in monitoring the project is wasted. The very fact that the stream of reports continues, although many of us confidently expected that within two years of the original flareup the issue would be dead, is in itself evidence that the UFO's remain an important public relations, public morale, and intelligence problem."[8]

It was a propitious time for Hynek to be able to take a breather, as the scientific world was set to undergo a tectonic shift, and the strange jigsaw puzzle of Hynek's career had placed him in a position to rise to the top of the upheaval.

Donald Quarles, who had so eloquently summed up the findings of Special Report #14 in the October press release, was about to trigger that scientific upheaval. As part of the International Geophysical Year (IGY), a massive celebration of international cooperation in earth science research, the Eisenhower White House had announced that the United States would launch the first-ever artificial satellite into Earth orbit, and Quarles was being tapped to coordinate the effort.

With the selection in August 1955 of the U.S. Navy's Vanguard missile as the launch vehicle, Quarles's team set about preparing for a 1957 launch. Members of the International Council of Scientific Unions had set the dates of IGY, somewhat incongruently, as July 1, 1957, to December 31, 1958, giving the U.S. rocket men and women ample time to test their ship and prepare its scientific payload. None of them was the least bit concerned when the Soviet Union announced its intention to follow the United States into space during the eighteen-month IGY.

Under the direction of Harvard College Observatory director Dr. Fred Whipple—who, incidentally, had authored one of the *Collier's* space travel articles—the Long Playing Rocket (LPR) committee was assigned by Quarles to analyze and report on "geophysical possibilities, desired orbit, technical feasibility, controls, motor, manpower, timing, budget, cost estimates, and other subjects related to the launching of an artificial earth satellite."[9]

Ultimately the LPR recommended that a ten-pound object should be placed in orbit, and that it should be "approximately 20 inches in diameter and be painted white or have an otherwise highly reflecting surface."[10] According to Whipple, if one had a proper pair of binoculars, "such an object could be observed visually from the ground at twilight, when it would be the equivalent of a star of the sixth magnitude."[11]

That was the trick. Even if the United States succeeded in launching a rocket, getting that rocket all the way up to an orbital altitude of one hundred miles or more, and then releasing a satellite at a velocity of 18,000 miles an hour, how would anyone on Earth know that the satellite was actually up there? Whipple's statement that an object moving at 18,000 miles an hour could easily be spotted at twilight through ordinary binoculars may have been scientifically sound, but it challenged the imaginations of Americans, many of whom had trouble spotting a stationary redheaded woodpecker in their own backyards through their binoculars. Shouldn't such an exacting task be left to the country's professional astronomers?

Whipple's observatory, newly combined with the Smithsonian Astrophysical Observatory in Cambridge, Massachusetts, in 1955, provided the perfect facility from which to develop history's first satellite tracking project. When it was all but certain late that year that the IGY grant money would be coming his way, Whipple approached Hynek, the expert at observing unusual stars, about becoming the observatory's associate director in charge of the program.

Whipple's assistant director at the observatory at that time was another of Hynek's acquaintances, Dr. Donald Menzel. One can only imagine the conversations Whipple and Menzel must have had concerning Hynek's employment, as Menzel and Hynek had crossed swords over the UFO controversy at the meeting of the Optical Society of America in 1952.

In another twist, IGY itself was the brainchild of physicist Lloyd Berkner, a member of the infamous Robertson Panel that had so outraged Hynek with its crass dismissal of the Tremonton film and the UFO phenomenon in general. Berkner was, in fact, the scientist who showed up two days late and yet still felt qualified to offer his opinion on the topic of UFOs.

Whipple wisely placed a barrier between Hynek and Menzel: while Hynek would run the Optical Tracking Program, a job that would entail some serious globe-trotting, Menzel would stay put in Cambridge, managing the day-to-day operations of the observatory. There would be little, if any, interaction between the two men, both of whose talents Whipple obviously valued.

With little fanfare, OSU and Ohio Wesleyan granted Hynek

a leave of absence to work on the Vanguard program, and "Hynek embarked on this, the most exciting part of his career to date, although a colleague at Ohio [State] urged him: 'Don't go! They'll never get it up!'"[12]

One of Hynek's fondest wishes was about to become reality. The longed-for international union of scientists, gathered together in peacetime in the pursuit of a single grand purpose, was taking form. The International Geophysical Year was, in the words of its proud creator, Lloyd Berkner, "perhaps the most ambitious and at the same time the most successful co-operative enterprise ever undertaken by man,"[13] and Hynek, the new director of the Smithsonian's Optical Tracking Program, was to play a grand role in the effort.

Sixty-seven countries, "acting outside the usual political and diplomatic framework of nations,"[14] had committed resources and personnel to the global study of the planet Earth, its atmosphere, and its relationship with the sun. This study was aided by a period of heavy solar activity—an absolute blessing from the cosmos—that was to make the atmospheric and space research especially meaningful. It was a cornucopia for the sixty thousand participating physical scientists the world over, who, suddenly elevated to international celebrity status, shared their secrets with journalists, teachers, students, and laypeople around the globe and thus demystified our home planet and its place in the solar system.

Working with one of the preeminent astronomers of his day, and installed at one of the nation's premier scientific institutions and placed in charge of a high-profile program with global reach and a princely budget just north of $3.3 million, Hynek attacked the seemingly insurmountable technical and human challenges with zeal. Within months, he and Whipple had defined the primary components of the tracking network: on the technical end, the program required twelve tracking stations strategically placed around the globe, each equipped with a mobile camera to photograph the moving satellite against the background of fixed stars, timed by crystal clocks accurate to within one millisecond, and tied into a central data processing center in Cambridge; on the human end, the program needed personnel to man the twelve remote tracking sites and calculate the satellite's orbit, plus thousands of volunteer sky-watchers stationed in remote locations throughout the newly united world of IGY, to make initial visual contact with the speeding satellite.

The cameras themselves had not yet been invented, so Hynek's first task was to address that inconvenience. Modeling the ideal tracking camera on a Super-Schmidt unit previously used by Whipple in a Harvard study to photograph meteors, Hynek found that the Schmidt's 12.25-inch aperture, f/.65 shutter speed, 52-degree field of view, and simple pivoting base were not adequate to the task at hand.

To capture images of the first satellite, Hynek brought together optical engineer and astronomical camera designer Dr. James Baker and mechanical engineer Joseph Nunn. Baker modified the Schmidt design to incorporate a sixteen-inch aperture, f/1 shutter speed, a 20-degree field of view, and a modified film transport. When the navy abruptly announced that its Vanguard satellite would have to be reduced in size from a thirty-inch diameter to twenty inches in order to fit in the rocket, the new camera had to be modified accordingly to capture an image of a much smaller object. Baker's emergency upgrades included a twenty-inch aperture, a reduced image diameter, and an expanded 30-degree field of view.

The mechanized mount created by Nunn, meanwhile, needed to alternate between a stationary photography mode and a tracking mode each time the satellite passed overhead, a complicated proposition. Not surprising, then, that when the photographic assembly was connected to the movable mount, the Baker-Nunn camera measured a monstrous twelve feet high and twelve feet across.

"The manufacture of the camera was one of the proudest achievements of American industry," claimed the Smithsonian, noting that there was no time for a prototype or for testing: "Once the large components were put into production, there was no opportunity to change any of the details."[15]

To make the most of the cameras' abilities, Hynek needed

them installed in locations that would allow for the very best satellite viewing. Accordingly, tracking sites needed to be roughly between the 30th north and south parallels, in locations with minimal cloud cover and with unimpeded views of the horizon. And even in this moment of international unity, they needed to be located in cooperative countries.

Hynek's thought, in accordance with the spirit of IGY, was to bypass political boundaries and take advantage of the "closely-knit fraternity"[16] of astronomers, making arrangements directly with his colleagues in other countries. The governing board of

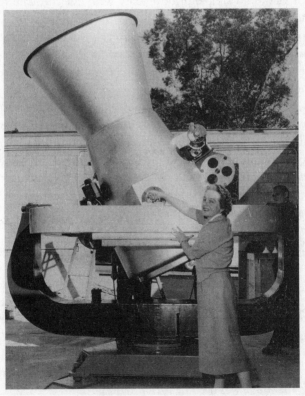

Mrs. Joseph Nunn demonstrates the ease with which an operator can adjust the Baker-Nunn camera, codesigned by her husband. A dozen of these cameras were positioned around the globe and formed the backbone of Dr. Hynek's Optical Tracking System. *(Courtesy of Northwestern University)*

America's IGY program, however, thought it best to work through the State Department, which ultimately cleared the way for the construction of tracking bases in Australia, the Union of South Africa, Spain, Japan, India, Peru, Iran, Netherlands Antilles, and Argentina. Additional domestic sites were established in Hawaii, New Mexico, and Florida.

As the Optical Tracking Program was gearing up, Hynek made two notable hiring decisions. In January 1957, he brought on Andrew "Bud" Ledwith III, a talented radio engineer with a background in UFO investigations. The following month he hired a young student named Walter Webb as an office assistant. Both men soon found themselves invited to the Hynek household in Belmont, Massachusetts, to help their boss evaluate UFO case reports telexed from the Project Blue Book offices in Ohio.

Together Hynek, Ledwith, and Webb formed a hush-hush UFO investigative team, analyzing case reports from Blue Book after hours and keeping mum about it at the tracking project offices. When Webb proposed to write an article for a national UFO newsletter, Hynek cautioned him that if such an article was ever seen by a higher-up in the IGY national committee—or, worse yet, Whipple himself—they might think that Hynek was "hiring flying-saucer enthusiasts instead of trained personnel."[17]

FOR ALL ITS WONDROUS OPTICAL POWERS, the Baker-Nunn camera had one glaring weakness: it could not be used to actually *locate* the Vanguard satellite. That had to be carried out by the weakest link: the human factor.

This was not a bug, but a feature. Hynek and Whipple had always known that before the tracking stations could follow the satellite through the sky, that satellite would first have to be located by ground observers at several scattered locations. From those observations, the satellite's orbit would be calculated and its location and trajectory wired to the tracking stations. Only then would the station operators know where to point their Baker-Nunn cameras.

It was Hynek's extreme good fortune that he had volunteers lining up to help find the world's first artificial moon.

"[IGY] aroused a remarkable enthusiasm among the peoples of many countries," wrote Smithsonian Astrophysical Observatory staff writer E. Nelson Hayes. "Millions saw in it an example of international goodwill and cooperation such as they had only too rarely known. The IGY, particularly its space projects, fired the imagination of people who, when called upon to do so, gave to it their fullest cooperation."[18]

Enter Moonwatch, a mighty all-volunteer army of amateur stargazers, numbering more than eight thousand at its peak, who could think of no more enjoyable way to spend their nights than methodically surveying the skies and, with luck, catching the first sight of the first unnatural object in space. "In many communities, Moonwatch took up where Chautauqua and similar activities of the 19th century left off," Hayes wrote. He later added, "Moonwatch drew people away from such passivity and back into a community activity in which many could participate either directly or indirectly. Even those who were not members of a local Moonwatch team could derive much satisfaction from supporting it."[19]

Thus, Moonwatch became the first international crowdsourcing project. It was the kind of cooperative scientific project that Hynek had discussed with Otto Struve years earlier, but on a more massive scale than Hynek might ever have imagined.

The Moonwatchers would dutifully, cheerfully, gather at twilight and again at dawn, when the satellite would be most visible, and set up their telescopes—some purchased at scientific stores, but many homebuilt—in a "fence." Each observer was a "picket" on that fence, and when an observer spotted the satellite he or she would get a reading on its location and report it to the group timekeeper. That information would be radioed to Cambridge, where it would be correlated with sightings by other teams.

When the project was announced in early 1956, the response was immense. "Visual observation teams sprang up in North America, South America, Africa, Europe, and Asia, in the Middle East and at such remote specks on the map as Station C and Fletcher's Ice Island T-3 in the Arctic Basin . . . Teams were organized by universities, high schools, government agencies, commercial organizations, private science clubs, and groups of laymen."[20]

And so, eighteen months after the initial plans had been laid, all the pieces were in place for Project Vanguard to shake the world to its core.

At 6:30 on the evening of Friday, October 4, 1957, the offices of the Optical Tracking Program at Harvard's Kittredge Hall were nearly empty. Only Hynek and his assistant Kenneth Drummond had yet to leave. As they idly discussed their plans for the weekend, the phone rang and Hynek picked up.

At first he thought it must be a prank call. Had the voice on the other end just said that a *satellite* had been launched? By the *Russians*? The reporter on the phone persisted, and when Hynek asked him to read the full wire dispatch he knew it must be true.

"Only then did I realize that this really was IT," Hynek was to tell his staff a year later. "As I hung up the receiver I just said to Ken in I'm afraid a rather weak voice, 'They've done it. There is a Russian satellite up.'" Sputnik had been launched.

"For a moment or two, I can tell you, both Ken and I felt extremely helpless," he said. But then the phone rang again, and adrenaline kicked in. While Hynek handled calls from reporters, Drummond started phoning the staff, telling them the news and ordering them back to the office. Many who heard the bulletin driving home simply turned around and headed back to Harvard.

"This was indeed the zero hour," Hynek said, "not the one we had planned for, but here nonetheless."

Nothing and no one was in the right place. Fred Whipple was in Washington attending an IGY conference and wouldn't be getting back to Boston until later that night. The only operational Baker-Nunn camera was at the manufacturer's, in Pasadena, California, where it had been disassembled for an overhaul. And there was no operating teletypewriter machine in the communications room at Kittredge Hall; in fact, there was no communications room—it had never been set up.

And reporters were starting to arrive.

Kittredge Hall was soon besieged by reporters and radio and TV crews, while staffers returned to their posts throughout the night and into the morning. Webb and Ledwith rigged receivers to pick up Sputnik's beeps, the first artificial signals from space. "I remember staying until 4 a.m.," Webb wrote, "to listen in fascination to the weird chirp of the Soviet satellite as it passed overhead every 96 minutes—the sound that captivated the world and shocked the United States into an educational and scientific revolution."[21]

If Kittredge Hall settled into a mystified silence every ninety-six minutes as the assembled hordes listened to Sputnik, it descended into chaos during the intervening ninety-five-minute periods. "My office was soon a shambles of extension cords, cigarette butts, and used-up flashbulbs," Hynek recalled,[22] while the glare of TV and movie lights gave Kittredge a brilliant halo that alarmed the neighbors. "It must have been a spectacular display," Hayes wrote, "for a woman living several blocks away reported that the building was on fire, and soon confusion was compounded by a pumper and a hook-and-ladder dispatched to the scene."[23]

"During the first few nights of Sputnik I, Kittredge was alive with people completely around the 24 hours," Hynek said. "Yes, even at three and four a.m. there was the buzz of activity." Everyone on staff gave up sleep and their weekend to help push out satellite data and predictions. Mrs. Hynek and Mrs. Whipple delivered "gobs" of coffee and sandwiches around the clock, and even clean shirts when a change was needed.

"Before long . . . we had a world-wide nerve center focused in one room on the second floor at Kittredge," Hynek said, and from there he and Whipple assured the country that there was no imminent threat from a Soviet attack. As the Moonwatch teams scrambled into action and the Baker-Nunn camera was speedily reassembled, Hynek and Whipple held twice-daily press conferences, hiding nothing, sharing everything they knew and becoming the faces and voices of calm and reassurance to millions of worried Americans.

"The slightest word of Whipple or Hynek to the press carried enormous weight," reported Hayes.[24] Because the satellite's orbit

Shortly after Sputnik's launch, Hynek (*right*) appeared on a news special produced by Boston TV station WBZ, assuring viewers that he and his colleagues were keeping a close eye on the Soviet device. *(Courtesy of Northwestern University)*

made it impossible to sight over North America for the first few days, the two astronomers could make only their best guesses as to Sputnik's position, based on the scant information being released by the Soviets. "Their guesses turned out to be correct, and this fact helped to establish in the minds of both the press and the general public that the word of Observatory officials was reliable."[25]

Four days after Sputnik's launch, the first confirmed Moonwatch sighting was made by teams in Sydney and Woomera, Australia. Two days later, on October 10, the first confirmed sighting in the United States was made by the Moonwatch team in New Haven, Connecticut. Because the Baker-Nunn cameras would not be fully deployed and operational for another five months, Hynek and Whipple were entirely dependent on their Moonwatch observers for tracking information; the teams exceeded expectations.

"They were there when the shooting started," Hynek said of his Moonwatch volunteers, thanking them "for having met the very vital challenge that the world's first Sputnik launched into our personal skies."

Within two weeks of the launch, Hynek and Whipple shared the cover of *Life* magazine; they are shown plotting Sputnik's orbit around a colossal globe—so big that Hynek stands on a stepladder to plot the orbit over the Arctic—and the stark headline simply reads: "The Satellite: Why Reds Got It First, What Happens Next."

What many Americans felt *should* happen next was that the United States should match the Russians and get its first satellite off the ground, pronto. But the Vanguard rocket proved to be uncooperative; after three successful suborbital test launches in the summer and fall of 1957, the first Vanguard intended to launch an American satellite into orbit faltered only four feet off the launchpad. The rocket crashed back to Earth and exploded into a humiliating ball of flame. The Soviets had, by this time—December 1957—effortlessly launched two satellites and one dog into orbit, and the United States couldn't get more than four feet off the ground.

The trauma to the American psyche would take a very long time to heal.

"It is only in retrospect now that one can sense what really happened," Hynek said on the first anniversary of the Sputnik launch. "The public reaction to Sputnik was, as you recall, a strange mixture of awe, admiration and fear, the latter enhanced of course because there had been no warning. As my barber put it, 'Most people realize that things in the sky were put there by God, and here was something that was brand new in the skies which although it wasn't put there by God, must have been put there by someone awfully close to God.'"

CHAPTER 9

INTERACTION

GOD, OR PERHAPS SOMEONE NOT CLOSE TO HIM AT ALL, put a second Sputnik into orbit on November 2. This one carried a dog, a forlorn little mutt named Laika, who gave her life for science, perishing miserably after only four orbits.

Millions of people went outside to watch for the second artificial satellite in the sky, and perhaps hear Laika's plaintive barks, but one man in Levelland, Texas, would have to wait for another night to watch and listen. Patrolman A. J. Fowler was the duty officer at the Levelland Police Department that night, and even if he had wanted to duck outside for a peek at the sky, the steady ringing of the phone would not have allowed it.

The first of Officer Fowler's many odd calls came in at 10:50 P.M., just an hour after the launch of Sputnik 2. The voice on the line was strained, scared, and uncertain; the man wanted to report a UFO sighting. "I was traveling north and west on route 116, driving my truck. At about four miles out of Levelland, I saw a big flame, to my right front," the witness said in his report, "I thought it was lightning. But when this object had reach to my position [*sic*] it was different, because it put my truck motor out and lights. Then I stop, got out, and took a look, but it was so rapid and quite some heat that I had to hit the ground. It also had colors—yellow, white—and it looked like a torpedo, about 200 feet long, moving at about 600 to 800 miles an hour."

The caller was a farm laborer and part-time barber named Pedro Saucedo, and he was accompanied that night by a friend, Joe Salaz. When Saucedo hit the dirt, Salaz remained in the cab as the object flew over them, rocking the truck. Salaz backed up Saucedo's claim that the truck's lights came back on after the object had passed, and that Saucedo was able to start the engine right up afterward. "Although Saucedo sounded terrified," Hynek was to report later, "the officer on duty did not at that time take the report seriously."[1]

That started to look like a mistake an hour later, when Fowler received a second UFO report. This call was from a Jim Wheeler, who claimed to have been traveling four miles east of Levelland when "he came upon a brilliantly lit egg-shaped object, about 200 feet long, sitting in the middle of the road. As [Wheeler] approached it, his car engine failed, and the headlights went out."[2] Officer Fowler could not fail to notice that Wheeler's object was in the area that Saucedo and Salaz had claimed their object had jetted off to.

"According to the observer, the object was lit up like a large neon light and cast a bright glare over the entire area," Hynek noted. "The observer decided to get out of his car, but when he did so, the UFO rose and, at an altitude of about 200 feet, the object's light or glare blinked out immediately. Mr. W. then had no trouble starting his car."[3]

The third report came in only minutes later, again from a motorist at a phone booth. This caller, Jose Alvarez, had come across a large glowing object sitting in the road about eleven miles north of Levelland. His car suffered a complete electrical failure, but then came back to life the instant the object's lights went out.

Fowler's phone rang again at 12:15 A.M. A motorist named Frank Williams reported seeing a glowing object near where Alvarez had had his encounter, only this time the glow pulsated, and with each pulse, Williams's car died. This went on until the object "left with a thunderous sound."[4]

"By this time Officer Fowler had finally realized that something odd was going on," Hynek observed drily, "and he notified the sheriff and his colleagues on duty, some of whom took to the roads to investigate. Two of them reported bright lights, seen for just a few seconds, but they did not have any car-stopping encounters."[5] Dispatchers from neighboring communities, meanwhile, were listening in on Fowler's communications and having a good laugh at the patrolman's expense. "Finally, though, when Fowler called for reserving the air for emergency messages only, the kidding ceased. Before the night was over, everyone involved was serious."[6]

The next sighting was called in from a pay phone at 1:15, this time from "a terrified truck driver."[7] The trucker reported to Fowler that his lights and engine failed as his truck approached a brilliant, glowing egglike object that gave off an intermittent neon glow. He claimed that the object was about two hundred feet long and gave out a roar when it took off straight up into the sky. "Officer Fowler stated that the truck driver was extremely excited when he called and that the witness was most upset . . ." Hynek said. "The truck engine and lights worked perfectly when the object left."[8]

Levelland fire marshal Ray Jones, who had been listening in on Fowler's dispatches, was out in his car to look for the glowing UFO and got his wish: shortly after the truck driver's sighting, he "encountered an object and experienced brief light and engine difficulty."[9]

Reports continued to come in the next day. A Texas Tech student named Newell Wright reported to the police that he had been approaching Levelland at 12:05 A.M. the previous night "when he noticed his ammeter jump to discharge and back, then his motor quit as if it were out of gas and the lights went out. He got out and looked under the hood but could find nothing wrong. Turning around he saw on the road ahead an egg-shaped object with a flattened bottom, like a loaf of bread and glowing not as bright as neon. No portholes or propellers were visible.

"Frightened, Wright got back into his car and tried to start it, but without success. After a few minutes, the egg rose almost straight up, veered slightly to the north and disappeared from view in a 'split instant.' After it was gone, the car started normally."[10] Another sighting reported to the police the next day switched up the pattern. This time the witness saw the glowing red-orange object approach from a distance and land on the highway a quarter mile away, upon which its glow changed to blue green. This encounter had taken place at 12:45 A.M. the previous night, not far from where Saucedo and Salaz saw the first object just two hours earlier. "The witness reported that the motor of the truck he was driving 'conked out' and his headlights died," Hynek noted. "Meanwhile, the object sat there on the road ahead of him, glowing bright enough to light up the cab of his truck."[11] After a minute, the object made a vertical ascent, its lights reverting to red orange as it did so.

"Officer Fowler reported that a total of 15 phone calls were made to the police station in direct reference to the UFO," Hynek wrote. Fowler noted that all fifteen callers were "very excited" about what they had encountered, and rightly so: when an American is prevented from using his or her vehicle—one of our most potent symbols of personal power and liberty—to either confront or escape imminent danger, something has gone terribly, fundamentally wrong with the world. What's more, Levelland sits on a huge swath of uninterrupted flatland in the Texas Panhandle, a landscape that does not provide any natural cover; if you're caught on a highway near Levelland, there is no place outside your car short of the ditch to run and hide.

In addition to rendering cars and trucks inoperable at the most disturbing moment possible in the middle of a flat, barren nowhere, the Levelland UFO—dubbed "the Thing" by witness Alvarez and "Whatnik" by a clever reporter—was notable in that it was seen close at hand by a demographically diverse group of witnesses in a concentrated geographical area, all in the span of a few hours.

The possibility that the Levelland occurrences were coincidental "is out of the statistical universe," Hynek later wrote.[12] He also dismissed the notion of mass hysteria—which would not have explained the electrical failure of so many vehicles in any case—or the idea that Saucedo's sighting somehow made it onto the news and was heard on the car radios of multiple drivers in the area that night; after all, Fowler hadn't taken Saucedo's call seriously and would not have been likely to report it to the media. All that Hynek had left was "prevarication," a most unlikely explanation for the night's events.

According to Hynek, Project Blue Book paid shockingly little attention to the Levelland sightings, which had made national headlines: "I was told that the Blue Book investigation consisted of the appearance of one man in civilian clothes at the sheriff's office at about 11:45 A.M. on November 5; he made two auto excursions during the day and then told Sheriff Clem that he was finished."[13]

"Captain Gregory, then head of Blue Book, did call me by phone," Hynek recalled of the Levelland case, "but at that time, as the person directly responsible for the tracking of the new Russian satellite, I was on a virtual around-the-clock duty and was unable to give it any attention whatever. I am not proud today that I hastily concurred in Captain Gregory's evaluation as 'ball lightning' on the basis of information that an electrical storm had been in progress in the Levelland area at the time. That was shown not to be the case . . .

"Besides, had I given it any thought whatever, I would soon have recognized the absence of any evidence that ball lightning can stop cars and put out headlights."[14]

AS THE UFO PHENOMENON QUIETED DOWN in the late 1950s, and Project Blue Book managed to avoid making headlines, some followers of the scene read the apparent inactivity as surefire evidence of a government cover-up. To fill the vacuum, a new UFO "supergroup" coalesced in 1956, dedicated to establishing a nationwide reporting system for UFO cases and a scientific approach to studying those reports.

The new group called itself the National Investigations Committee on Aerial Phenomenon (NICAP) and soon attracted high-profile board members such as UFO author Donald Keyhoe. The group appointed Keyhoe as its director in 1957, and he set out to reignite and transform the country's conversation about UFOs. Because his antigovernment stance and insistence that UFOs represented an extraterrestrial phenomenon had sold so many books over the years, Keyhoe saw no reason to change his approach with NICAP.

The Levelland events were a case in point. Keyhoe and his group wasted no time in looking into the goings-on of that night and made the most of the enthusiastic press coverage. "That's just the way an electromagnetic force field would operate," a confident Keyhoe explained to a reporter about why so many cars and trucks had died around Levelland. "This isn't the first time we've heard of such things happening."[15]

Keyhoe's nearest NICAP investigator was James Lee, the proprietor of a hardware store in Abilene. Dispatched to interview the Levelland witnesses, Lee quickly decided that this could be the Big One, the case that changed everything. "It's an amazing, fantastic story," he told a reporter. "Those people out there saw something—there's no doubt about it. It was literally something out of this world.

"Whatever this thing was, it had a definite mission and a plan of operation."[16]

ON JANUARY 31, 1958, with nearly a full year left in IGY, the United States launched its first artificial satellite, Explorer 1, into Earth orbit. Instead of launching a dog into space, the United States had focused on packing scientific instruments into its first satellite, similar to the payloads Hynek had designed for the V-2s launched from White Sands a few years earlier.

Two days after it was launched, Moonwatch teams in Bryan, Texas, and Albuquerque, New Mexico, spotted the American satellite, and its orbit was calculated from their data. Soon after, it was spotted by trackers and photographed by Baker-Nunn cameras in South Africa, Japan, and New Mexico. Hynek's Optical Tracking

Program, only a notion in 1955, was, three years later, hitting all its targets. In fact, by the time IGY was ending, the program was exceeding all expectations, tracking eight U.S. satellites and three Soviet "birds."[17] Once envisioned as a finite operation that would end with IGY, America's satellite tracking program had become a permanent, and irreplaceable, component in the United States' plans to conquer and dominate space.

When the National Aeronautics and Space Administration (NASA) was created by an act of Congress in July 1958, the Smithsonian Astrophysical Observatory gradually withdrew from the satellite tracking business, ceding primary responsibilities to the new civilian agency. It had been a good run; the techniques, equipment, and human network conceived and created by Hynek set the standard for tracking man-made objects in Earth orbit and set a curious milestone for science: Hynek had been tasked with developing a means of observing and measuring something that did not yet exist, and was considered by many to be impossible, and somehow he succeeded.

CHAPTER 10

OBSCURING INFLUENCE

OVER TIME, the Sputnik age began to have a strange effect on the UFO phenomenon.

First, there were the photos. In the wake of Sputniks 1 and 2 and Explorer, more cameras than ever before in human history were pointed at the skies, and it was perhaps inevitable that at some point or another something unusual would appear on a photo taken by one of them.

Walter Webb, fresh from setting up a radio receiver in Kittredge Hall to listen to Sputnik's infernal beeping, was assigned to man the Optical Tracking station on Maui, Hawaii. This station, high atop Mount Haleakala, had been intended to be the last of the twelve to begin operations, but in the post-Sputnik scramble it became the first to go online. This put Webb at the center of tracking activities for a time, and he was aware that observation teams were noticing things that they weren't actually looking for. He reported in 1993 that Bud Ledwith and Allen Hynek had begun noticing "oddities" in both tracking station film and Moonwatch reports concerning "strange lights that certainly weren't satellites."[1]

On January 24, 1958, for example, a Moonwatch team in Fort Worth, Texas, reported seeing a "White object; Seems to merge with *Gamma Geminorum*." On August 20, a Moonwatch team in Roanoke, Virginia, reported a "bright, flashing round object seen 3 times." On September 8, 1959, a Moonwatch team in East Rock, Connecticut, spotted an "Oblong object of 'tremendous size'; Bright." On April 14, 1959, the Optical Tracking station in Mitaka, Tokyo, sighted a "Flashing object (1/3 sec. frequency)" and captured photos of same.

"I certainly know that in the satellite tracking mission we got a number of things that appeared on the films that were never tracked down; they weren't part of the mission!" Hynek had confessed to Webb. "A person who says the Baker-Nunn cameras never picked up anything is just dead wrong because I know they did. I was in charge of the project!"[2]

Because more people were watching the skies for artificial objects—not really knowing what these objects might look like—more people were seeing UFOs. But the more people were seeing UFOs, the more UFO sightings could be dismissed as misidentifications of Sputnik.

Such was the case when Kenneth Houston saw a UFO on October 18, 1958, in Papua New Guinea, and mentioned the sighting to his friend, a skeptical Anglican priest named William Gill. Father Gill was unimpressed and suggested to his friend Houston that "he had seen the Soviet satellite Sputnik."[3]

Gill would have cause to reconsider his comment the following June, when UFOs appeared in the sky on three consecutive nights near the Anglican mission he led in Boianai, Papua New Guinea.

Father Gill sighted the first object—a large, four-legged, hovering disc with a blue light beam shooting intermittently up from its rim—at about 6:45 P.M. on the night of June 26, 1959. He estimated it to be equivalent in size to five full moons in a line and reported that it maintained an altitude of about 150 meters, under a cloud ceiling of about 600 meters.

Four humanoid beings came out on the top "deck" of the object, "busy at some unknown task,"[4] and apparently completely uninterested in Gill and the others who had gathered with him on the beach. In Gill's notes, he described the strange activity the best he could: "One object on top moves—man? Now three men, moving, glowing, doing something on deck. Gone."[5] Years later Gill was more precise about the glow, telling Hynek that the figures were "illuminated by the blue light, as well as by an 'aura' that surrounded them and the craft without touching them."[6]

Nearly forty mystified witnesses, including Gill, his staff, and dozens of natives, watched the glowing occupants until the object was obscured by cloud cover some forty-five minutes after it appeared. Gill noted that the light of the disc "reflected like a large halo in the cloud."[7] Clearly, this was not Sputnik.

A short time later, some smaller objects became visible, after which the "mother ship," so named by Gill, reappeared. The assorted objects would remain visible until nearly eleven P.M. "Gill thought maybe it was 'the Americans,' but if it was the Americans it was pretty bloody clever," explained Bill Chalker, an Australian UFO researcher and historian, and longtime friend of Father Gill's. Because he didn't want to give himself a chance to change his mind in the morning, Gill had the presence of mind to write up and then sign a statement that night attesting to what he had seen. Many of the other observers felt the same way and signed on to Gill's statement.

The following night, the mother ship reappeared along with two smaller objects. Again, four glowing figures were visible on the top of the larger object, and again, they were absorbed in some mystery activity hidden from people on the beach. This time, however, Father Gill wasn't nearly as sure of what he was witnessing. "It sure didn't look like Americans," Chalker said. "Who are these guys? They're glowing and, if it was as close as they thought it was, these figures seemed to be bigger than six feet. To him that was very impressive."

One of the impressive figures held on to a sort of ship's railing and seemed to be watching the dozen or so observers on the ground. Certain that the figure was interested in them, Father Gill had a thought: he raised his arm in a wave. "To our surprise," Gill said, "the figure did the same."[8]

One of Gill's companions waved with both arms, and two

of the figures on the object waved both arms back. Gill and his friend waved together, and in response all four of the mysterious visitors waved back. "There seemed to be no doubt that our movements were answered . . ."[9]

One of the people on the beach flashed the beam of a flashlight at the disc, and, in what appeared to be a response, the disc swung

In one of the strangest "occupant" cases on record, an Anglican missionary in Papua New Guinea, his staff, and local villagers observed large humanoid figures atop a floating disc for two nights. Witnesses reported that when they waved their arms in greeting, the humanoid occupants waved back. *(Artwork by Allan Hendry. Courtesy of the J. Allen Hynek Center for UFO Studies.)*

from side to side, like a pendulum. After several more exchanges, the ship approached the humans on the ground, but according to Chalker there was no sense of menace, only excitement. Gill and the others had been watching the object and its strange occupants so long, they were anxious to see and understand what they were looking at.

But the ship didn't land. After several seconds it stopped and retreated, to the annoyance of Gill and the others. "They fully expected it to come down and land, and all that ambiguity would have been taken away at that point," Chalker said.

The third night played out differently from the first two. As many as eight objects were visible at one time, but there were no figures on hand; whatever task they had been busy with the two previous evenings must have reached its completion. No one seems to have waited around to see the last of the objects off. The magical moment, whatever it might have represented, had passed.

"Critics of this case do not generally know that [Father Gill's] report is only one of some 60 in the New Guinea area at approximately that time, all investigated by a colleague of Gill, the Reverend Norman Cruttwell," Hynek was to write later.[10] Cruttwell's report made its way to several civilian UFO groups in Australia, and, buoyed by the news that Father Gill would soon be moving to Melbourne, those groups showered the Australian Parliament with copies of the report and demands that an official investigation be undertaken.

In time, the Australian air force looked into the Father Gill case and "could come to no definite conclusion of the report." This did not stop it, however, from deciding, Blue Book–style, that the objects were most probably "reflections on a cloud of a major light source of unknown origin."[11] This crude attempt to bury the Gill event in a shallow grave only served to keep it vibrantly alive.

Hynek first learned of the Gill case from the British Air Ministry during a 1960 Blue Book visit. The Brits did not take the case seriously and were all too happy to pass the case file on to Hynek: "It had apparently been cluttering up their files."[12] The Air Ministry would have gladly handed over all its UFO files to Hynek, in fact, as it had suspended all UFO investigation activity secure in the knowledge that the Yanks were on the case and that Blue Book would soon solve the mystery.

The Gill case, which now boasted New Guinean, Australian, British, and American involvement, underscored the fact that UFOs were coming to be seen as an international problem. UFOs had been reported in more than 140 countries, and there was confoundingly little variety in the types of objects reported: what witnesses described seeing in the skies above Rio were essentially identical to what was being reported by witnesses in Turkey, Canada, France, and Japan, as though they were all manufactured on the same UFO assembly line.

Hynek was troubled that the phenomenon was seeming more and more to affect the entire human race in a surprisingly consistent manner, but he took strength from the fact that scientists and governments around the globe were at least paying attention to UFOs, even if some of them were sitting back and waiting for the U.S. Air Force to solve the problem for them.

It would be well over a decade before Hynek would have an opportunity to look into the Gill case in any depth, and in one respect that delay was crucial. In 1960, Hynek was deeply ambivalent about UFO reports involving occupants, and he was, in all likelihood, just as uncomfortable with the case as were his British colleagues. "Back in the day of Blue Book, he probably would have accepted it as a throwaway—it had to be astronomical," Chalker said. "But clearly, his view evolved through time, as the Boianai case kept coming up."

Where did Hynek belong? The simple answer in 1959 was *anywhere he wanted to be,* as no one could have been riding higher: "Project Moonwatch was a great success, both scientifically and popularly," said author Ian Ridpath, "and Hynek became something of a celebrity."[13]

The stars were not leading him; he was leading the stars. As he approached his fiftieth birthday, and the satellite tracking system he spearheaded was absorbed into NASA's operations, several paths lay open to him. He could follow the tracking program to NASA and be a leading light in the space program. He could remain at Harvard and continue to work under Whipple. Or he could do something that would give him almost limitless autonomy and near anonymity.

"After [the Optical Tracking System] was in running order, I accepted the offer of chairmanship of Astronomy at Northwestern," Hynek reported in notes for an undated speech. "It was a difficult choice, whether to remain at Smithsonian-Harvard, or to come to Northwestern. I chose the latter mainly because of the greater opportunity to do research work of my own choosing."

"I came here to work with you on S.T.P. on January 1, 1956, when satellites were just a gleam in the scientific eye," Hynek wrote to Fred Whipple in his letter of resignation. "Now they are commonplace, and we are tracking them with success. Time clearly for me to turn to more challenging things!"[14]

In his handwritten reply, Whipple revealed the true extent of Hynek's influence at the Smithsonian Astrophysical Observatory:

> *Your leaving SAO is a real personal loss to me as you not only were a real friend—but an optimistic, positive person who was always eager to do something new. Never could I have lived through some of our vicissitudes without your continued encouragement and backing.*
>
> *We all miss you and Mimi more than you will know or believe. Thanks for investing 4½ years with me.*
>
> *Regards ever,*
> *Fred[15]*

A return to teaching at a university may seem a surprising—almost disappointing—choice for someone seemingly on the brink of greatness, but after several years of backbreaking work and incessant pressure, Hynek needed to breathe. "Even though he was so close to the start of it, Hynek still admits himself dazed by the pace of the space age," observed Ridpath.[16] Rather than continue to track objects that race through the sky every ninety-six minutes, Hynek chose to return to the comparative peace and quiet of tracking objects that lope along through the heavens at their leisure and then don't return for days, weeks, or months . . . As a clear sign of Hynek's growing need to return to the more sedate pace of nature's astronomical clock, he embarked on a trip to Madrid, Spain, in July 1959 to observe the mid-afternoon occultation of the bright star Regulus by the planet Venus. To say that this was a rare event is to underestimate both the precision of the universe and the patience of astronomers: this event, in which Regulus would be seen to pass behind the dark side, or limb, of Venus and reemerge from behind the illuminated limb a little over eleven minutes later, had never before been observed in recorded history.

It got to Hynek.

"It is not customary today in scientific accounts to dwell upon the beauty of a phenomenon. I would leave break with this custom to remark that the moments just before the occultation presented a spectacle of singular beauty," he wrote in *Science* magazine after the event. "If one considers the familiar crescent moon-and-star symbol as represented on a flag . . . but now replaces it by a miniature, bright yellow, crescent moon with an actively twinkling greenish-yellow star almost within the crescent, and both set as living jewels in a clear turquoise matrix, one will, I hope, pardon my digression."[17]

Around the same time Hynek took another trip to witness a full solar eclipse, but this time he traveled up, some forty thousand feet, to see the spectacle from a military jet. Although this lifted him above the interfering clouds, it did not protect him from a replacement pilot mistakenly flying a course that put all Hynek's instruments on the wrong side of the plane. "Still," he remarked somewhat philosophically, "I was one of the few people who saw the eclipse at all that year since it was raining on the ground."[18]

Hynek's move to Northwestern coincided with a personal baby boom, as new sons Paul and Ross were born in the first years of the 1960s, just as Allen and Mimi's three older children were already in their late teens and early twenties and venturing off to college and careers. Having two little ones underfoot after coming so close to being empty nesters was a sign that Hynek felt more financially secure in his new post than he had ever felt in academia.

Although Hynek's new position at Northwestern University began on January 1, 1960, he asked for and received an eight-month leave of absence so that he could finish up his work in Cambridge. Two projects in particular had been launched at the Smithsonian Astrophysical Observatory in the wake of the satellite tracking project, and they demanded Hynek's ongoing management and problem-solving talents. Both projects reflected Hynek's longtime interest in capturing better images of astronomical objects, and both had deep roots in his past.

"Telescopes, in effect, are anchored at the bottom of a vast and turbulent ocean of atmosphere, an ocean which distorts and blurs incoming signals from the stars," Hynek explained in an official Northwestern biography. "The atmosphere completely blocks entrance of information-packed energy in the short ultra-violet, X-ray and gamma ray regions, and large portions of the infra-red and radio regions as well. This is as frustrating to astronomers as a radio which can only tune in one station even though dozens are broadcasting."[19]

To further complicate the life of an astronomer, constantly shifting layers of varying air temperatures at different altitudes can bend and distort starlight, as can heat escaping from under an observatory dome in the dead of winter and swirling around the end of an icy telescope. As Hynek had learned at Yerkes and Perkins in his youth, the best images were gotten when the interior of the observatory was kept at the same temperature as the outdoors, even on a frigid January night in Wisconsin or Ohio.

Project Star Gazer, an attempt to transcend that interference by raising an astronomical observatory above the atmospheric ocean, was first discussed at a November 1957 meeting among Hynek, Whipple, and Major David Simons, chief of the U.S. Air Force's Space Biology Branch.

After surveying the methods and technologies available to them in 1957, the three men decided that "a forceful program of

astronomical observations from balloons is in order." Hynek proposed that a high-powered telescope could be sent as high as thirty
thousand feet in a manned balloon, during which time the pilot
would take "a one minute recording of scintillation on at least
three bright stars of various elevation angles above the horizon."[20]

If one of those words resonated, it's because Hynek saw this
balloon project as a natural, inevitable progression of the work he
had done under government contract while at Ohio State to study
the scintillation of stars. That experiment had derived in turn
from an earlier project Hynek had undertaken for the air force
to develop methods of detecting and photographing stars in the
daytime. Through those earlier experiments, Hynek had discovered that stellar scintillations occurred at very high altitudes; that
image motion, "the primary bane of astronomical observations,"[21]
was a phenomenon that occurred close to the telescope lens; and
that certain binary stars scintillated separately even while their
images moved in unison. He also found that scintillation varied
throughout a twenty-four-hour cycle and, perhaps strangest of all,
that wind speed and direction aloft could be determined by observing the scintillation of stars. Thus, "with the study of stellar
scintillation and stellar image motion carried to great lengths from
ground observations it was obvious that the next step was to add
the vertical dimension to this study," Hynek reported.[22]

Not long after that first meeting, Hynek shared an even grander
vision, one that would ultimately shape much of the United States'
space program for decades. In a television news interview that
aired in 1958, Hynek urged the construction of a "National Space
Observatory" by the proposed national space agency. Claiming
that a space-based observatory would "pay great and immediate
dividends" in new scientific knowledge, Hynek "pictured the observatory located in the U.S. with a satellite station orbiting the
earth, high above the earth's atmosphere."[23]

"From a space observatory we could see the surface of planets
with unimaginable clarity even with a small telescope," Hynek
said. In addition to revolutionizing astronomical observing,
Hynek's proposed telescope could also be trained on Earth itself,

making real-time weather data "continuously available to weather forecasters over the world" and leading to "greater knowledge of basic weather causes" that would result in more reliable long-range forecasts.[24]

Hynek was, of course, describing the Hubble Space Telescope a mere thirty-two years before its launch, while throwing in a little preview of TIROS-1, the first successful weather satellite, to be launched two years hence, in 1960. Few scientists can claim to have predicted such an earth-shattering advance in human knowledge over three decades in advance, but to Hynek it was all in a day's daydreaming.

THE IMAGE ORTHICON PROJECT, also begun at the Smithsonian Astrophysical Observatory, sought to counteract the effects of the atmospheric ocean in a different way. Hynek had always been fascinated by television, so it was perhaps natural that in his quest to identify ever-fainter objects in the sky he would revisit his near adventure with the photoelectric cell and flashlight under the dome at Yerkes Observatory in 1933, and consider the amazing light-gathering properties of a TV camera. According to Ridpath, "At one of the Baker-Nunn sites—Las Cruces, in the New Mexico desert—Hynek and Whipple had experimented with a television camera attached to a telescope. From this has developed what Hynek calls 'image-orthicon astronomy,' a branch of observing that has grown vastly in popularity since Hynek helped pioneer it."[25]

"In no field of science is the battle to make every photon count more intense than it is in astronomy, where light levels are notoriously low, so low indeed that one can even at times speak of the number of seconds per photon rather than the number of photons per second."[26] With that explanation, Hynek and two of his Northwestern colleagues, in a 1972 paper, underscored the value of electronically amplifying the brightness of distant astronomical images. Using a television tube, Hynek was able to amplify light captured by a telescope by a factor of up to one hundred and correspondingly increase the speed at which an image of a faint object could be developed.

"Electronic image conversion," the formal name for the system, became what the National Science Foundation described at the time as the greatest development in astronomy since photography,[27] and what Hynek described as his greatest contribution to astronomy.[28] In his optimism, Hynek made no secret of the fact that he would someday perform "a high-altitude marriage,"[29] by launching an Image Orthicon telescope in a Project Star Gazer balloon.

AROUND THE TIME THAT HYNEK was packing up and moving the troops to Evanston, the alien invasion tropes that had leaped from movie screens with alarming regularity throughout the 1950s had found a way to infiltrate American households. If you had a TV in your living room at the dawn of the 1960s—and who didn't?— you and your family would never be safe.

Two weekly anthology series set the tone with regular tales of fear and paranoia set in worlds where menacing aliens could be found in the most unlikely places and where space travel always turned out badly. And both series reeled audiences in with irresistible hooks delivered by ominous, disembodied voices: *"You open this door with the key of imagination . . ."* said one. *"There is nothing wrong with your television set . . ."* said the other.

The Twilight Zone, hosted by writer-producer Rod Serling, was the first to appear on living room TVs, beginning in late 1959. Although the nightmare stories of *The Twilight Zone* were often propelled by fate, fortune, and outright fantasy, some of the most jarring episodes dealt with devastating alien invasions or space-flights gone horrifically awry. Those include such classics as "The Monsters Are Due on Maple Street," an exercise in suburban paranoia in which any house on the block could be harboring invading aliens, and "The Invaders," in which an old woman in an isolated farmhouse battles off tiny aliens—who turn out to be Earth astronauts exploring a planet of giants. Perhaps best known and best loved is the episode based on a short story by Damon Knight, "To Serve Man," involving seemingly benevolent aliens who have come to Earth *to serve man,* only not in the way it first appears.

"That *The Twilight Zone* is damn near immortal is something I will not argue with," wrote author Stephen King in his critique of the horror genre, *Danse Macabre,* but he had to admit that the beloved show often veered from "moral tales, many of them smarmy," to stories that were "simplistic and almost painfully corny." Because of this, King said, "for sheer hard-edged clarity of concept, *The Twilight Zone* really could not match *The Outer Limits.*"[30]

The Outer Limits was a dark monster-of-the-week series that began each installment with an ominous command that viewers "sit quietly" as some unseen, unknown "we" took control of the TV transmission and forced them to "participate in a great adventure." When it debuted in 1962, *The Outer Limits* quickly out-aliened *The Twilight Zone,* as the show's producer, Joseph Stefano, fresh from writing Alfred Hitchcock's *Psycho,* dictated that every episode feature a "bear," or monster, that must appear by midpoint in the story. To make good on the threat of the opening narration, the premiere episode actually did portray an alien who transmitted itself through space and emerged on Earth from an experimental TV set built by a very surprised electronics engineer. In what would become the central theme of the show, the earthlings turned out to be the true monsters, while the blazing, radioactive alien, seeking only peace and understanding, proved far more humane than the humans.

Over its brief forty-nine-episode run, *The Outer Limits* told manifold stories of unsettling alien encounters, displaying a deep affinity to alien invasion and abduction tales, including one in which menacing aliens abduct six whole city blocks from an Earth neighborhood. Those nightmarish alien "bears" took the form of human-faced ants, surveillance-minded cyclopes, batlike humanoids, and, surpassing all others, the mute man-turned-alien "Thetan," from the first-season episode "The Architects of Fear," who was so terrifying that many ABC network affiliates blacked out viewers' screens when the monster appeared, virtually editing out the entire final act. They truly were in control of the transmission.

"Not being involved in science fiction, I approached it, perhaps fortuitously, from a very fresh viewpoint," said writer-producer Stefano, "and I have always felt that it is so arrogant of us to think that (A) we're the only planet with life on it and (B) that if there is life on other planets it's going to resemble us. I always thought that's perfectly insane . . . I thought there was a great chance for excitement and drama in the fact that another planet *would* have ants that had faces reminiscent of ours."[31]

An episode of *The Outer Limits* has been involved, quite improbably, in a UFO controversy that has simmered for decades. Some UFO critics have insisted that the aliens described in the first major American UFO abduction case, one that took place on a lonely highway in New Hampshire in September 1961, were first seen by the abductees in the form of the "light being" in the *Outer Limits* episode "The Bellero Shield."

Of course, if the abduction took place in September 1961, and the TV show didn't debut until September 1963, the timeline can't possibly work, can it?

Well, yes, it can, because the abductees, Barney and Betty Hill, did not consciously remember their abduction experience until after they began hypnotherapy treatments on February 22, 1964,[32] twelve days after "The Bellero Shield" had aired on ABC.

The Hills, a biracial couple from Portsmouth, New Hampshire, had been on a postponed honeymoon to Niagara Falls that week but had decided to return home early. As they drove south along an empty stretch of Route 3 that wove through the White Mountains, Betty saw a strange star in the sky near the moon. She pointed it out to Barney and remarked that it seemed to be getting larger. Barney soon saw it, too, and drew the most logical conclusion: "Betty, that's a satellite," he said to his wife,[33] and returned his attention to the twisting highway.

Barney pulled over so their dachshund, Delsey, could relieve herself, and Betty retrieved her binoculars to get a better look at the light. It was definitely growing brighter, and it was definitely moving. Barney was still determined to explain it away, now guessing that it was a Piper Cub heading north to Montreal, but

Betty wasn't so sure. Her sister had seen a UFO once, and Betty was open to speculating about flying saucers, even though her husband was a scoffer.

When they stopped to look again, near a scenic spot in the mountains named Indian Head, neither one was certain what they were looking at. Betty had seen the object pass in front of the moon, revealing a cigar shape, and it seemed to be ringed with flashing colored lights. As the silent object drew closer, they could see two rows of illuminated windows along its front edge. When it appeared to deploy wings on either side with red lights at the tip of each wing, Barney finally gave up on the Piper Cub idea.

Betty was starting to get upset, which bothered Barney no end. Betty was always the steady one; she wasn't supposed to be afraid! Barney looked north and south along the highway, thinking how nice it would be if the state police happened along sometime soon.

Now armed with the pistol he kept in the trunk of the car, Barney walked into a field where the object seemed to be headed. It descended over the field in a great arc, tilting forward toward the ground, and Barney found himself looking through the binoculars at a row of humanoid figures lined up along the row of windows. All but one of the figures stepped back, and the one remaining stared at Barney with its huge slanting eyes. "Don't be afraid. Stay there—and keep looking," it communicated to Barney.[34] Barney tried to look away but found he could not tear the binoculars from his eyes.

Back in the idling Chevy, Betty was clutching Delsey and screaming to Barney to come back to the car. From where she sat looking out the open driver's door, her husband had simply disappeared into the night, and she was starting to panic.

After several minutes Barney came tearing back to the car, in what Betty described as "a hysterical condition."[35] He leaped into the driver's seat as Betty and Delsey made room for him. Betty said, "Why didn't you come back? I was screaming for you to come back." "They're going to capture us!" Barney kept repeating as he jammed the car into gear and pulled out as quickly as the Chevy could move.[36]

Betty looked for the object in the sky but could see nothing above the car, not even the stars. It took a moment for her to understand that the object was over them, blotting out the night sky. Then they both heard a series of beeps coming from somewhere behind them, perhaps from the trunk, and things got foggy for a moment.

The beeps sounded a second time, and both Barney and Betty emerged from a sort of stupor. They were still driving south on

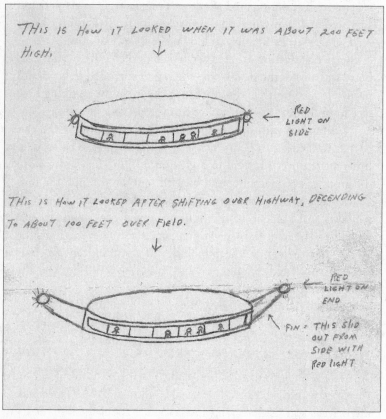

Barney Hill's UFO: First he thought it was a satellite, then a Piper Cub, and then, through the binoculars, it appeared as a winged disc with telepathic humanoid occupants visible through the windows. (*Artwork by Barney Hill. Courtesy of Kathleen Marden.*)

Route 3, dawn was approaching, and the encounter at Indian Head with the strange flying object seemed somehow distant.

As they headed home, Betty said with a giggle, "Do you believe in flying saucers now?"

"Oh, Betty," Barney said, "don't be ridiculous. Of course I don't."[37]

CHAPTER 11

BURNED BRUSH

"**DID CREATURE FROM SPACE BURN GRASS?**" asked the headline of an April 26, 1964, United Press International story. After that emphatic tease, the uncredited reporter led with a calmer, but no less provocative, statement: "It was round, stood on girder-like legs, and scorched grass and bushes, but Socorro Policeman Lonnie Zamora does not know what it was."[1]

Poor Zamora. All he wanted to do on the evening of Friday, April 24, was to hand out his quota of traffic tickets and then spend a quiet night at home with his wife. There was never much going on in the small desert town of Socorro, New Mexico, where Zamora lived and worked, but the town was ringed by popular vacation areas and busy government installations, like the sprawling White Sands Missile Range and Holloman Air Force Base. Because of this, a lot of drivers were just passing through, and some didn't pay heed to the local speed limits. When he saw a black Chevy racing past the courthouse toward the rodeo grounds, Zamora thought he might make his quota early and be able to punch out on schedule. At approximately 5:45, he took off in pursuit.

That's when he heard the roar and saw a flash of flame, perhaps three-quarters of a mile to the southwest, and decided to drop the pursuit. Fearing that the mayor's old dynamite shack in the hills might have exploded, Zamora veered off the road and onto a steep uphill gravel track.

After some struggle, Zamora got his car to the crest of the hill, from which he could see what he took to be the source of the commotion: "Suddenly noted a shiny type object to south . . . At first glance, stopped," Zamora stated. "It looked, at first, like a car turned upside down. Thought some kids might have turned over. Saw two people in white coveralls very close to the object. One of these persons seemed to turn and look straight at my car and seemed startled—seemed to jump quickly somewhat"[2]—a reaction that's sure to draw a policeman's interest.

As Zamora drove along a ridge, he lost sight of the object and the two figures, whom he noted seemed the size of "small adults or large kids" in relation to a nearby greasewood bush.[3] When he saw the object again, at the bottom of a shallow arroyo, the figures were gone. Zamora parked and radioed the station to say he was leaving his car to investigate a possible accident. As he descended the rocky slope, he could see that the object was not an overturned car, but he couldn't for the life of him say what it was.

It was a white aluminum oval, aligned horizontally and perched three or four feet off the ground on two outward-slanting legs. The two figures must have gotten inside, but how? The object was completely smooth, with no apparent windows or doors. He heard several thumps, akin to the sound of a car door being shut, then the roar started up again without warning. As flames erupted from under the object, Zamora scrambled back up the slope in fear. He bumped into the rear of his car, knocking his glasses off, but kept running. When he turned back, shielding his face with his arms, the object was aloft, quickly flying over the dynamite shack and off into the distance, the roar changing to a shrill whine and then silence.

"The object seemed to lift up slowly, and to 'get small' in the distance very fast . . ." Zamora reported. "It disappeared as it went over the mountain. It had no flame whatsoever as it was traveling over the ground, and made no smoke or noise."[4]

Zamora radioed for assistance, then drew a sketch of the red symbol he had seen on the side of the object: an inverted V with embellishments, "like a typical cattleman's brand," Hynek later noted.[5] Zamora then returned to the arroyo, where the object had stood a few moments earlier. "I noted the brush was burning in several places," he said.[6]

When Sergeant Sam Chavez of the state police arrived a few minutes later, he found Zamora sweating and as pale as a sheet. "[Chavez] had never seen anyone quite as scared as Lonnie and

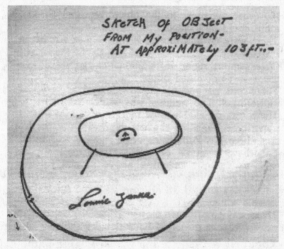

SKETCH OF OBJECT FROM MY POSITION-AT APPROXIMATELY 103 FT...

Lonnie Zamora

Patrolman Lonnie Zamora's sketch of the white aluminum craft he saw in a rocky arroyo near Socorro, New Mexico. Two humanoid occupants in white coveralls entered the object and flew away moments after they saw Zamora approaching in his squad car. *(Sketch by Lonnie Zamora, from the Project Blue Book archives)*

asked him, 'What's the matter, Lonnie, you look as though you've seen the devil?'"[7] Mumbling that he might have, at that, Zamora led Chavez to the landing site. Chavez saw the burning brush and found it still hot, and he saw something more: distinct depressions in the soil, indicating that something very heavy had been standing there on landing struts.

The response to Zamora's sighting was swift and decisive.

In a few hours he was being interviewed by Arthur Byrnes, an FBI agent from the regional office, and Richard Holder, an army captain and up-range commander from White Sands. After getting Zamora's no-nonsense statement, Byrnes and Holder joined Zamora at the site of the incident, where Socorro Police and State Police were combing the area for forensic evidence. "They confirmed that 'something' had been there and the Army took soil samples of 4" by 5" impressions which were found where the alleged vehicle had landed. The impressions appeared to be burned and the surrounding area was clean."[8]

After examining the evidence and taking measurements, Byrnes and Holder returned to the station with Zamora, only to learn from the police dispatcher that three people had called in reports of seeing "a blue flare of light in the area" where Zamora saw the object.[9] It would be learned later that a vacationing family had been buzzed in their station wagon by a low-flying object moments before seeing Zamora's car leave the main highway in pursuit. The convincing nature of Zamora's experience was quickly snowballing.

The news took a few days to hit the local paper, but that was only because the Socorro *El Defensor Chieftain* came out just twice a week. When the April 28 edition went to print four days after the incident, the story got an impressive amount of ink. "Zamora said he saw lettering on the side of the UFO . . ." the paper said. "He did not believe the lettering was in English and observed no numerals as there are on known aircraft. Zamora said he was not at liberty to further describe the lettering."[10]

After relating the story of the tourists whose station wagon had been buzzed, the reporter made a not-unreasonable declaration:

> *Zamora doesn't know what the object was, but for those who desire to speculate, there are three possibilities:*
> *First, it may have been a top secret U.S. aircraft in an advanced stage of development.*
> *Second, it may have been an advanced type of aircraft or space ship of another power.*
> *Third, it may have been a space scout ship from another planet.*[11]

The U.S. Air Force's Aerial Phenomena Branch had changed leadership yet again just a few months before the Socorro event. New project chief Captain Hector Quintanilla was, in Hynek's words, "the present incumbent at Blue Book."[12] A sworn enemy of the UFO, Quintanilla had recently quit smoking but lit up again when Lonnie Zamora's story set the newswires abuzz. He wanted all UFOs to go away, and he didn't appreciate it when a case like Zamora's came to his desk by way of both the Pentagon and the FBI and forced him to act.

Quintanilla's initial reaction to Zamora's report was dismissive. He suspected that Zamora had seen a classified military test vehicle, possibly even a prototype lunar lander, that had ventured too far from either White Sands or Holloman. Captain Holder, however, was quick to deflect Quintanilla's theory, telling the local paper that "neither White Sands Missile Range nor Holloman had an object that would compare to the object described."[13]

"The Air Force is in a spot over Socorro," said Hynek in confidential remarks recorded a few weeks after the event. "A vague statement identifying it as an unspecified U.S. experimental aircraft won't go down. Congressional inquiries have been received, and Quintanilla is under pressure for an answer." Unfortunately, Hynek lamented, "the Air Force doesn't know what science is."[14]

Four days after the sighting, Hynek was abruptly instructed by Quintanilla to join up with an analyst, Task Sergeant David Moody, and a UFO investigator, Major William Connor, and he flew from Chicago to Albuquerque on only three hours' notice. Hynek's orders were to gather as many facts as he could "before the incident becomes legend."[15] He was met at the airport by Major Connor, "a pompous chap,"[16] and ushered into a staff car for the seventy-five-mile drive south to Socorro. Along the way the car had a flat tire, and when he learned that there was no jack in the vehicle, Hynek decided, in what was to become a most appropriate metaphor for his work with the air force, to "thumb a ride" the rest of the way.[17] Ultimately, Hynek beat Connor and his car to Socorro.

Hynek quickly distanced himself from Connor and Moody

when he learned that Lonnie Zamora and Sam Chavez were, after four days of dealing with ongoing official bluster and equivocating, both "very anti–Air Force."[18]

"I got rid of the Air Force people and got the story from them that night at the jail," Hynek said, noting that both men were reluctant to talk. "Zamora is an unimaginative cop of an old Socorro family, incapable of hoax, and pretty sore at being regarded as a romancer. It took at least a half hour to thaw him out."[19]

Once thawed, however, Zamora went over the sequence of events one more time, impressing Hynek with his attention to detail and his thorough, unembellished account of his experience. In short, Hynek found Zamora almost abnormally credible. "The guy doesn't drink, cavort with women, or recite poetry," Hynek said. "He captures speeders."[20]

When Hynek investigated the site of Zamora's encounter the following day, he found that Coral and Jim Lorenzen of APRO had already come and gone. They had met with both Lonnie Zamora and Sam Chavez on Sunday, April 26, after driving overnight from their Phoenix, Arizona, headquarters. Their take on Zamora was that he "gives one the impression of quiet dependability."[21] He was, however, seemingly intimidated by alleged instructions from the air force not to discuss the landing pad imprints or the odd symbol on the side of the object. Although Major Connor denied that Zamora had been warned not to discuss the symbol, the fact that two different descriptions of the red mark eventually emerged—one with three horizontal stripes and one with an arch over a single horizontal stripe—lends credence to the Lorenzens' telling.

In any event, after interviewing the witnesses and inspecting the arroyo where Zamora had seen the object, Coral Lorenzen was certain of what had happened and what it meant: "I myself left Socorro convinced, for the first time in 17 years of UFO investigation, that alien beings are reconnoitering our planet."[22]

Hynek was, as always, far more circumspect. "It is one of the soundest, best substantiated reports as far as it goes," he told an AP reporter. "Usually one finds many contradictions or omissions in these reports. But Mr. Zamora's story is simply told, certainly without any intent to perpetrate a hoax. The story of course was told by a man who obviously was frightened badly by what he did see. He certainly must have seen something."[23] However, when asked by a reporter for the local radio station if he thought the object was a craft from outer space, Hynek merely replied, "You know about as much about it as I do."[24]

In truth, Hynek was bothered by the fact that there were no reports of the object appearing on radar, especially since that part of New Mexico was "infested with radar equipment."[25] He also found it odd that the object did not kick up a great cloud of dust upon takeoff. To make matters worse, by the time Hynek was able to investigate the arroyo, the burned brush was gone and the imprints in the soil had been trampled out of existence. "There's nothing there to see now," he complained.[26]

Both the military and the civilian UFO investigators were happy to address Hynek's concerns. The military explained the lack of radar readings by suggesting that Zamora's object never ascended to an altitude where it could have risen above the ground clutter and been picked up by radar. The civilian UFO investigators explained the lack of a dust cloud by suggesting that the object "most likely used a power source completely unknown to us."[27]

"I ended up rather doubting Lonnie's report," said Hynek's colleague and confidant William "Bill" Powers. "Not his report, but his imagination. He saw these people in jumpsuits outside of the spacecraft, and they got back in and left. But I could not decide if that was real."

Powers had joined Hynek's staff at the Dearborn Observatory on the Northwestern campus in 1960. He was studying behavioral psychology and control systems at the time and was looking for part-time work to pay his college bills. He found out that the observatory needed someone to "count heads" at the free Friday night public viewings through the main telescope, and a perk of the job was a free place to live in the back room of the building. Powers took the gig, but his facility with electronics made him an invaluable resource for the astronomy department, and in time he dropped out of the university and took a full-time staff position with Hynek. By the time of the Zamora case, Hynek was regularly bringing his trusted assistant along to help investigate cases for Project Blue Book and sometimes sending him into the field solo when his teaching schedule interfered with UFO investigations. This arrangement was not without its complications, as the independent-minded Powers could be very skeptical of what he encountered in the field.

When he visited Socorro at Hynek's invitation, Powers found Zamora's story challenging: "Either it was an air force test of something like a lunar lander, which would have worked as an explanation, but on the other hand, if you interpreted his descriptions, which was that they were of small stature and acted funny, then these were not human beings. Well, that put me to the test, really, about my attitudes. So I decided, all right, I'll just report on what I noticed and give it an honest analysis, and give it a chance to be real.

"It turns out that's what Hynek was doing, too. He also was putting his skepticism aside and trying to give it a fair shot," Powers said. "If it's really happening, we want to know, because that's something of considerable interest."

In short order, Zamora provided Powers with something of considerable interest. The policeman showed Powers a sketch of the "landing marks," with all measurements, angles, and distances carefully laid out. "And it suddenly hit me," Powers said, "that there was a geometric relationship here that would have put the center of gravity of this so-called UFO right over the main burn mark. I thought, 'Uh-oh, that's interesting.' Because that's exactly where a rocket would have to be, or a jet would have to be, to lift the thing off the ground."

It was observations such as this one that made Powers's talents so crucial to Hynek's investigations throughout the 1960s, even if Powers never quite knew on whose dime he was traveling. "I don't know where all his dimes came from," he quipped.

Hynek would admit later that he almost wished he could declare the case a hoax or hallucination, but Zamora was just too upright for him to consider it. Zamora's first thought after the sighting, Hynek said, was to ask his superior if he should talk to his priest. "His second was that he resented the whole thing because it prevented him from getting his quota of speeders that day!"[28]

"Despite my strong desire to find a natural explanation for the sighting . . . ," Hynek wrote, "I could find none; the case is therefore listed in the Blue Book files as 'Unidentified.'"[29]

"I think this case may be the 'Rosetta stone,'" he concluded. "There's never been a strong case with so unimpeachable a witness."[30]

IN SEPTEMBER 1963, the dynamic duo of Hynek and Powers welcomed doctoral student Jacques Vallee to the team at the Dearborn Observatory. Vallee had recently moved his young family from France to Austin, Texas, where he was to study astronomy, but he soon felt that astronomy was too stale and static a field for his boundless curiosity. Computer programming was the new frontier, and Northwestern University had a program that suited Vallee's sensibilities.

Northwestern also was home to Hynek, whom Vallee had long admired. Having been deeply involved in UFO research in his native country, Vallee was well aware of the significant contributions Hynek had been making to the field; to most Americans Hynek was still the Sputnik man, but to a growing community of people curious about strange aerial phenomena, Hynek was a person of much deeper and more mysterious interest.

When Vallee first met with Hynek in Chicago, their initial conversation lasted not hours but days, according to Vallee. "We had so much to talk about!" he wrote in his journal. "He is a warm and yet deeply scholarly man, with much energy and a great sense of humor, an open mind, and a deep sense of culture that comes from the sophistication of his Czech ancestry. [I] am impressed by his sharp ideas and his eagerness for action. He has a lively face where piercing eyes are softened by a little goatee that makes it hard to take him completely seriously."[31]

The small Romanesque building housing the Dearborn

Observatory—itself a sort of three-fifths scale version of the Yerkes Observatory seventy miles to the northwest—was bursting with expansive ideas and revolutionary thinking, and as Hynek showed him around the facilities and introduced him to Bill Powers, Vallee was hooked.*

"I want to work closely with Hynek," he gushed. "There was an immediate bond between us . . . He has a genuine understanding of science but his sense of culture saves him from the pitfalls of specialization. His house in Evanston is full of books on all kinds of subjects. There are classical records everywhere, and current issues of the cartoon-filled *New Yorker* on the coffee table. He is an ethical man, with a realistic view of the politics of science and the role of the military. After our very first meeting he tried to call [Blue Book project chief] Colonel Friend to suggest that he join us in Evanston right away. I like this impulse; it comes from a man of action who can make quick judgments about people."[32]

Before long, Vallee was educating Hynek on the progress made in France by science writer and UFOlogist Aimé Michel, making the case that patterns could be identified in UFO events, and encouraging Hynek to develop a computer database of Blue Book's thousands of UFO case files.

The injection of new energy couldn't have come at a better time for Hynek, who had his hands full running the astronomy department and raising funds for not one but two new observatories, even as he continued to pour energy into the Image Orthicon project and Project Star Gazer. A solar eclipse in the summer of 1963 had given Hynek a rare opportunity to put the Image Orthicon system into practice in the field, and he mobilized the entire astronomy department to do so.

Hynek's primary goal was to capture the first color photographs and movie film of a total solar eclipse.

---

* Powers and Vallee had something besides an interest in UFOs in common: both men had published science fiction stories as well, Vallee in France and Powers in American pulp magazines like *Galaxy* and *Astounding*.

After suffering through two massively frustrating eclipse missions in the past, he couldn't have wished for better luck on his mission to Canada. Not only was his party one of the choice few to be in the right place to see the eclipse that summer, but so spectacular were the photos Hynek brought back that *Life* magazine ran a special full-color pictorial of the eclipse images in its August 2, 1963, issue. What's more, Hynek got a shot of the flash spectrum, the first color photo of the phenomenon ever taken.

It was a very good thing that the Image Orthicon system was proving to be so versatile and useful, because Hynek's other pet project was about to come crashing down around him.

Early indications showed nothing but promise for Project Star Gazer, and Hynek had had no trouble signing up enthusiastic sponsors who recognized the simple genius and scientific merit of lifting a powerful telescope above the atmosphere in a balloon. He had received full funding for the project from the air force and had secured the use of a state-of-the-art pressurized gondola, built by the air force for high-altitude medical testing. He had acquired the most advanced balloon available and had recruited a fearless high-altitude crew: air force balloon pilot Captain Joseph Kittinger Jr., who then held the world's record for the highest successful parachute jump, and U.S. Navy astronomer and first-time balloon passenger William White, a former graduate student of Hynek's from Ohio Wesleyan University (and, almost too charming to be believed, former babysitter of the Hynek children). A series of four manned launches was scheduled to take place at Holloman Air Force Base outside Socorro, New Mexico, with an expectation that "the first one or two flights might experience difficulties."[33]

That they did: on the first test flight, in December 1962, White and Kittinger achieved the target altitude, an astonishing 15.5 miles above the earth, but White was unable to get the balky tracking mechanism to lock on to stars. The malfunction—caused by a damaged electrical amplifier—made stellar observations and imaging all but impossible, and White was unable to get any useful data. The second test, the following March, was

The Project Star Gazer balloon rises from the New Mexico desert. The two-man crew in the pressurized gondola would ride 15.5 miles up to capture images of celestial objects from above the atmosphere. *(Courtesy of Northwestern University)*

scrubbed before liftoff; a premature firing of the balloon's release mechanism—likely triggered by static electricity caused by high winds—resulted in the balloon coming loose from the gondola and getting tangled in the launch tower.

By the time of the third launch, a month later, anticipation was high. Hynek, Kittinger, and White were confident that the technical glitches were all solved, and they would soon have proof that Star Gazer worked. But static electricity once more caused the explosive squibs to fire prematurely, this time before the balloon had even been attached to the gondola. The balloon didn't simply come loose; it was carried by prevailing winds some one hundred miles to the east-northeast, all the way to a little town called Roswell. Once the balloon crashed to the ground, "the wind dragged it around so that it was rendered completely useless for any flight."[34]

The loss of the balloon was too big a blow. After this third failed attempt, the air force notified Hynek that it would not be funding the final planned flight. The many organizations and institutions that had supported the project "were rudely shocked, particularly the balloon astronomy team at Northwestern."[35]

"Rudely shocked" is putting it nicely. Hynek was incensed. Never before had he experienced such an abrupt, unreasonable withdrawal of support on the very brink of success in such a promising experiment; not even his frustrations with Project Blue Book had ever inspired such a towering rage. So furious was he that it took him over three years to submit the final report on Project Star Gazer, and even after that lengthy cooling-off period he was still sizzling with anger. "The setting aside of a project which had engaged so many for such a length of time, at a time when success seemed assured, can only be listed in the scientific annals as a criminal act, and one carried out in a callous, cavalier manner without regard for the desires, objectives and ideals of the people involved," Hynek wrote. He was particularly upset and guilt-ridden over the fact that White, a former student and a personal friend, "gave four years of his scientific life which might well have been spent in other more productive directions."[36]

"In one fell swoop," Hynek closed his report, "the Air Force shot down any hopes the Northwestern team had for getting a just return on the many years of scientific effort they had pumped into Star Gazer."[37]

As he had done so many times in his career, Hynek overcame the disappointment and set his sights higher: this time, on the moon and Mars.

When President John F. Kennedy announced his intentions in May 1961 to land an American on the moon by the end of the decade, funding for space science projects spewed forth from the NASA coffers like the flames from a rocket nozzle. The United States was in a race with the Soviet Union to launch as many satellites and space probes as it could, with undisputed space supremacy going to the first to conquer the moon. Cost was no consideration.

Kennedy's assassination in November 1963 did nothing to slow the NASA steamroller; if anything, his death infused the agency with a new sense of urgency, and Hynek was not hesitant to hitch a ride. A continual flow of ambitious proposals—some designed to further understanding of Earth's neighboring planets and some intended simply to track the growing number of planetary probes being launched at Cape Kennedy (then Cape Canaveral)—came out of the Dearborn Observatory. Many of Hynek's new ideas were, quite naturally, based on the proven Image Orthicon system, but some co-opted the tarnished but still promising concept of Project Star Gazer. The most impressive of these was Hynek's February 1965 proposal for a moon-based observatory called the Lunar Vacuum Observatory. The proposal, nicknamed LUVO, was for a precision-mounted, remote-controlled twelve-inch telescope to be left behind on the moon's surface by the Apollo astronauts, who were expected to be making regular lunar landings by the end of the decade.

"The Luvo telescope would be free from the hazing effect of the earth's atmosphere," read a May 3, 1965, newspaper report.[38] From its perch on the lunar surface, LUVO would be used to reveal new wonders on the other planets of the solar system, as well as on Earth itself, both on its surface and in its atmosphere. It would fulfill the mission for which Star Gazer had been designed, but would do it better. "If you had a telescope on the moon," Hynek said, "you could point it where you want to by remote control and the picture would go into an image orthicon to be flashed back to earth on the same laser beam that controls the moon telescope."[39]

Although the proposal was met with initial enthusiasm by NASA, the lunar telescope was not deemed a priority. LUVO wasn't rejected outright, just put off indefinitely. In the end, LUVO did not reach the moon, but Hynek made it there just the same. He continued to work with NASA throughout the 1960s, using the Image Orthicon at Corralitos Observatory to track Apollo moon missions and to help select lunar landing sites.

THE HUMAN RACE'S VIEW of its place in the universe was irrevocably altered in August 1965, when the NASA space probe Mariner 4 finished transmitting a series of twenty-one photographs (and a partial twenty-second image) to Earth. Mariner 4, a six-hundred-pound craft that measured 9.5 feet high and stretched 22.5 feet across with its four solar panels fully extended, was the most ambitious NASA mission of its time, the first space probe from Earth to perform a close flyby of Mars, coming within six thousand miles of the planet's surface.

It could be said that the people of Earth had been anticipating these photographs with a certain amount of repressed anxiety since Giovanni Schiaparelli discovered the mysterious channels, or *canali,* on Mars nearly ninety years earlier, for they revealed for the first time the true nature of the red planet.

Up until the moment Mariner 4 finished transmitting its images of a bleak, parched, frozen world to Earth, millions of humans still harbored the belief that Mars was populated by intelligent, perhaps malevolent, beings, and that the planet's surface was covered with irrigation canals bringing water from the ice caps to the crops the Martians needed to survive. In the course of one transmission from Mariner 4, that image disintegrated, and the fantastic proclamations of Shiaparelli, Percival Lowell, H. G. Wells, and T. J. Weems all surrendered every last bit of their substance.

It was a lonely shock, knowing we had the solar system to ourselves. "There was disappointment among some scientists, and the public alike," said one science writer.[40]

It also posed a problem for UFOlogists. If not from Mars, then from where, exactly, did those little men in the white coveralls come?

CHAPTER 12

WILL-O'-THE-WISP, PART ONE

FRANK MANNOR LOCKED THE DOORS to his house for the first time ever on the night of March 21, 1966. Before that night, he and his family never had any reason to feel unsafe or insecure in their rented farmhouse on a muddy dead-end road in Dexter Township, Michigan. The night of the twentieth changed all that.

Frank was home between trucking runs, spending a tranquil evening with his wife and children, when the family dogs started to raise a ruckus outside. "It was about 7:30 p.m. when the dogs suddenly went crazy. Even the cattle were raising a racket like I've never heard before," Mannor, age forty-seven, told a reporter. "Well I ran out on the porch to shut them up. The German Shepherd was just wild, pulling on his leash and howling out toward the field."[1] That's when he saw the strange lights shining out from the swamp to the north of the house.

The animals quieted down as Mannor watched the lights change from blue and white to red, bobbing up and down all the while. He called his family out onto the porch to see, and after a few minutes, a potent combination of nerves and curiosity got to Mannor, and he and his eighteen-year-old son decided to stalk the strange intruder.

"Me and Ronnie went down for a look," Mannor said. "My

wife, Leona, kicked up a fuss. Said we might get radioactive. I asked my two sons-in-law to come along, but my daughters wouldn't let them."[2]

Despite Leona's fuss, Frank and Ronald set out without flashlights or guns. "We started walking out across the field. Sneaky like. I've been a hunter and trapper for years, and I was stalking it. Slow, without lights."[3]

Back at the house, Leona decided to notify the Washtenaw County Sheriff's Department, and son-in-law Bob Wagner called in a report of an object that "went up in the air about 500 feet and came down, making a lot of noise."

By Frank's accounting, he and Ronald took a full half hour to creep within five hundred yards of the object, and his descriptions of what he saw through the trees differed with every telling. In his account to a local reporter, Frank described an object that looked like a pyramid with a rounded top, with a "coral—rugged—corrugated" skin, and blue, red, and white lights. But in his account to the *Chicago News* four days later, he said that while he "could not make out the form of the object through the swamp mist, [he] estimates that it was about the size of a car and was hovering about 8 or 10 feet off the ground."[4] In an article in the *New York Journal-American,* he described a rotating red light, like that on a police car, and in yet another account to the *Detroit News,* Mannor described a pyramid with a porthole.

What remained consistent was that the lights changed colors, and the glow rose and fell regularly. After watching the glow for a few minutes, Ronald exclaimed, "Look at that horrible thing!"

Washtenaw County sheriff's deputies Stanley McFadden and David Fitzpatrick arrived at the Mannor home at about nine P.M., where they took statements from Wagner and the rest of the family. Wagner reported that the object in the swamp was displaying lights that alternated from blue green to brilliant red to yellow, and that it "appeared to be having difficulty getting off the ground." The deputies drove with Wagner to a good observation point just north of the swamp, on Quigley Road. "While in the woods area, a brilliant light was observed from the far edge of the woods, and upon approaching, the light dimmed in brilliance," said the deputies' report. "A continued search of the area was conducted, through swamp, and high grass, with negative results."

McGuiness Road would never again be as quiet or as remote. "My wife says we'll move out of here," Frank Mannor grumbled. "She doesn't like that. I never lock the doors."[5]

By the time the incident was over, at least fifteen Dexter Township police officers and Washtenaw County sheriff's deputies had been dispatched to the scene or had shown up of their own accord with family members in tow. Any other time this might have seemed an oddity, but in March 1966 it was perfectly normal, as local law enforcement officers had been reporting—and giving chase to—unidentified lighted objects in the sky for nearly a week.

The sightings in Dexter had started on March 14, at 3:50 A.M., when Deputies Buford Bushroe and John Foster first reported strange objects in the sky while out on patrol. Their first call set off a two-and-a-half-hour chase that stretched across three counties and out over Lake Erie. Law enforcement officers from five area jurisdictions were swept up in the excitement, and even Selfridge Air Force Base reported that its tower was tracking some objects over the lake. Two days later, Deputy David Fitzpatrick— the same deputy who was to respond to Bob Wagner's call on the twentieth—announced that he had photographic proof of the UFO visitations, and soon his photo of two luminescent streaks in the early-morning sky above Milan, Michigan, was appearing in newspaper and TV reports across the country.

Later that week, Sergeant Neil Schneider and Deputy Fitzpatrick reported seeing multiple "top-shaped" UFOs in the sky while they were out on patrol. Their description was short on detail, only that the multihued lighting on the objects changed intensity when the objects changed speed.

After that frenetic week of sightings, the Mannor swamp incident seemed a spectacular culmination of Michigan's UFO adventure. But it was not even close to being over. The night after the Mannor sighting, on Monday, March 21, violent thunderstorms battered southeast Michigan, making sky watching all but impossible. Sixty miles from Dexter, the storms were making it hard for coeds in the New Women's Dormitory at Hillsdale College to wind down after a long day of exams.

Cynthia "Pinky" Poffenberger, age eighteen, had been sitting in her second-story room watching the storm when she first saw something out of place. "I was sitting on my bed with all the lights out, looking out my window," she reported. "Finally, the storm stopped. About five minutes later I noticed something coming across the sky from the left. It seemed to be moving faster than an iarplane [sic]."[6]

Then she screamed.

Barbara "Gidget" Kohn's eyewitness account, appearing in the March 24 issue of the *Hillsdale Collegian* student newspaper, is far more abrupt: "UFO! The scream echoed down the hall of the second floor east wing of the New Women's Dorm."[7]

Kohn ran to her window and saw an object heading straight for the dorm, "radiating intense silver-white light." In a flash of lightning she saw what appeared to be a squashed football or basketball, although she could not estimate its size or distance. The object appeared to dart "north, then south, then up and down.

"It seemed to be frantic," she wrote.[8]

A GROWING GROUP OF YOUNG WOMEN watched the lights settle in a hollow in the Slayton Arboretum, approximately 1,500 feet from the dormitory. That, of course, is 500 yards, the same distance from which Frank Mannor claimed to have made his closest observation.

In total, eighty-seven residents of the New Women's Dormitory saw the lights. The more witnesses who arrived, the more the descriptions of the lights proliferated:

"It was blinking a red light that would get very intense and then it would flash to a white light, such as the light on the top of a police cruiser," said Poffenberger.[9]

"There seemed to me to be a sort of high-frequency hum," said Kohn. "The shape was hard to tell . . . The whole thing was glowing with silver light. Later on we saw individual lights."[10]

"Jo [Wilson], meanwhile, was watching the lights in the Arb[oretum] very closely for she thought they were moving," Kohn reported. "She called Sara [Robechek] over to the window and Sara told her that she thought it was a house because the whole object was yellowish white and appeared to be floodlights shining on a house."[11]

While Kohn set about recording the testimony of her fellow dorm residents, Sara Robechek ran down the hall to the telephone and called the nearby airport. She got no answer. The airport was closed, no flights at all that night. Still, the airport's beacon was lit, and all witnesses agreed that every time the beacon swept in the direction of the arboretum, the glowing lights would dim out.

The next logical step was for the women to alert their housemother, Kelly Hearn, but she was out of the building counseling a student. At least some of the women decided that they were facing an emergency, and Hearn had taught them what to do in case of an emergency: contact the Hillsdale County civil defense director.

The wife of the civil defense director and local undertaker William "Bud" Van Horn took the message and told the women she would send her husband right over. In the meantime, Hearn had returned to the dorm and quickly learned what had been happening in her absence. "I was met by a small knot of girls who informed me they had sighted a UFO, had called Civil Defense, and were expecting the arrival of a representative from that organization," she reported.[12]

Before he drove to the college, Van Horn called the police, and a squad car soon arrived outside the dorm. "The car drove right by as though they hadn't seen the object," Kohn wrote, "but the object had seen the car because it started to act frantic. It went up and down and back and forth fast and jerkily as if it didn't know what to do."[13]

When he arrived at the dormitory, Van Horn accompanied Hearn and the girls up the stairs to room 204. Inside the unlit room, twenty girls were watching the lights in the arboretum.

"After my eyes had been allowed to become accustomed to the dark, I observed an arrangement of lights in the distance between the dorm and the airport beacon," Hearn reported. "If one took the two lights farthest from each other to be limit lights, they looked to be about 20 feet apart to my untrained eyes. Intensity of light varied so that it seemed a rheostat were dimming them, sometimes to nothingness, then turning them to fuller intensity. At the approach of any light whatsoever, the lights of the object would dim to nothingness."[14]

One of the girls fetched a pair of binoculars and Van Horn, now in charge of the scene, started to make his own observations, as reported by Kohn: "He announced at this time that it was definitely a UFO and that it had been proven so by the facts of its maneuverability, motion, and the fact that even when it was stationary it was not on the ground."[15] The next day Van Horn would elaborate, telling a UPI reporter that "it was definitely some kind of vehicle."[16]

Only Kohn remained at the window observing the lights as they vanished, reappeared, and receded from the arboretum throughout the rest of the night and into the next morning: "We continued to watch for our friend, for in a sense it had become our friend, and a few minutes later we were rewarded by a strange new light on the horizon which hadn't been there before—a bluish whitish greenish light," she wrote. "None of us ever saw it move past the [beacon] light, and the whole time we watched it, it appeared to become very erratic, anxious, and nervous when anything resembling lights got near it.

"Finally I saw it disappear," she wrote. "I had gotten out of bed again at about 5:10 a.m. and as I watched it move from the northeast to almost due east and then get smaller and smaller until it completely disappeared."[17]

Hynek was not altogether surprised by the growing furor in Michigan. The previous year had seen staggering increases in the volume of UFO sightings reported to Blue Book, and the whole country had had flying saucers on the brain for months.

When John G. Fuller, a longtime columnist for the *Saturday Review,* wasn't happy with the air force's explanations, he decided to investigate a UFO sighting himself. Like Hynek, Fuller came to the UFO phenomenon as a pure skeptic, so when he dedicated an October 1965 column to an unbiased report of a recent sighting in Exeter, New Hampshire, his readers took notice and a sensation was born.

The case involved a teenager who, while hitchhiking home from his girlfriend's house, had seen flashing lights in the woods and then a large metallic object that hovered in the sky. Two local policemen investigated his story and were themselves buzzed by the strange object. So impressed was Fuller by the witnesses' sincerity that he started to question his own skepticism, and he was soon writing regularly about extraterrestrial craft and government cover-ups in his popular column.

Fuller was quickly commissioned by G. P. Putnam to expand his Exeter column into a book, and as the publication date approached, tantalizing excerpts appeared in the February 22, 1966, issue of *Look* magazine. It was brilliant marketing, guaranteeing immense interest when the book, *Incident at Exeter,* hit the stores.

In this new climate, Hynek found that the air force was suddenly enamored with his idea of letting scientists take a look at the problem and offer some expert advice. Major General E. B. LeBailly took Hynek's request seriously, and in February 1966, optical physicist Dr. Brian O'Brien, a member of the Air Force Scientific Advisory Board, convened a special scientific committee to consider the UFO. The O'Brien Committee, which included in its numbers a young Cornell University astronomy professor named Carl Sagan, came to a conclusion that made everyone happy: it recommended ongoing study of the UFO problem and offered the air force a way to shift that responsibility onto someone else.

To conduct this research, the committee called for a network of universities to study the UFO phenomenon. Each university, Hynek explained, would "study perhaps a hundred sightings per year, devoting an average of ten man-days to each investigation and the resulting report."[18]

The air force was pleased with the idea of handing UFO study over to universities, and Hynek was pleased that UFO research would finally be released from the dingy confines of the Project

Blue Book offices at Wright-Patterson and be conducted by real scientists at respected institutions of learning.

It was a small, noncommittal step forward with which everyone could be comfortable, at least for now. But the Dexter-Hillsdale sightings were about to force Hynek and the air force to make a very big step that neither was ready to make.

"Allen Hynek probably thought he was finally getting somewhere," said UFO historians Michael Swords and Robert Powell. "And then he played his final fateful role in this process."[19]

THE PUBLIC WAS HUNGRY for a sequel to John Fuller's feature story in *Look* magazine when the Michigan sightings made the news a few weeks later. This new situation was quickly developing into a more sensational story than the Exeter event, which may explain why Major Quintanilla was hesitant to become involved at first.

Hynek sat in Evanston and read the news stories from Michigan with growing impatience, wondering why on earth Blue Book was not responding. He called Quintanilla and asked why he had not been sent to investigate, and the major, obtuse as ever, replied that no one had reported the sightings to Blue Book. When Hynek countered that Quintanilla was not being very scientific, the major replied, "I don't give a damn."[20]

Hynek hung up the phone in disgust and stewed. Something big was happening in Michigan, and it would be criminal to stay put in Evanston. The sightings should be investigated thoroughly while the witnesses' memories were still fresh and before physical evidence could be damaged or stolen. Before he could come up with a plan, the phone rang and Hynek heard Quintanilla's voice on the other end, asking him to catch the first available flight to Michigan.

Hynek was incredulous: "I thought you didn't give a damn. I thought it hadn't been reported."

"Well, since I told you that, it has been reported."

"By whom?"

"By the Pentagon, doggone it! All three networks are talking about nothing else. The brass is having a fit!"[21]

After finishing up his day's classes, Hynek met with Jacques Vallee and Bill Powers about his trip to Michigan. It was decided that Hynek would make the trip alone, but he would call for his colleagues to join him if he felt he would need help with the investigation.

Hynek could already tell from the news accounts that the Michigan sightings would likely present him with a decided lack of hard evidence, a prospect he did not relish. "When I received the first accounts of the UFO's, I recognized at once that my files held far better, more coherent and more articulate reports than these," he wrote some months later. "Even so, the incident was receiving such great attention in the press that I went to Michigan with the hope that here was a case that I could use to focus scientific attention on the UFO problem."[22]

Upon his arrival, Hynek traveled to Dexter, some eighty miles to the west-southwest of Selfridge AFB, to begin his investigation. His first stop was the Mannor farm, which was well into its second day as the Dexter area's biggest tourist attraction.

"When I arrived in Michigan, I soon discovered that the situation was so charged with emotion that it was impossible for me to do any really serious investigation," Hynek said. "The Air Force left me almost completely on my own, which meant that I sometimes had to fight my way through the clusters of reporters who were surrounding the key witnesses whom I had to interview."[23]

Over and over again, Hynek found himself at the mercy of the mob that had descended on the Mannor farm. The local law enforcement officers were so much a part of the story themselves that they offered little help in maintaining any sense of order.

"Once upon the scenes . . . I realized that the task was far too great for one man if all the witnesses wishing to be interviewed (and many of whom were 'miffed' because I had either not had a chance to talk with them or was able to give only a short time to them) were to be satisfied," Hynek wrote in his Blue Book report.

Hynek now faced pressure on several fronts: he knew that Quintanilla would be expecting a quick resolution to the case, and he knew that his colleagues back in Evanston would be expecting him to push back against Quintanilla and demand time and resources for a thorough investigation. It would have been a difficult feat to walk that tightrope under the best of conditions, but Hynek had recently fallen and broken his jaw; his mouth was wired shut, and he was not operating at full strength. "I think if I'd been in top-notch health, I would have been thinking straighter," he later confessed.[24]

"The papers, the radio, all the media are feverish . . ." wrote Vallee. "The military are worried. This is the first large series of sightings since Fuller's widely read article in *Look*. A radical change has taken place in public opinion, the situation is mature. Hynek should speak out."[25]

But Hynek was too absorbed to speak out, as the radical change in public opinion threatened to smother his investigation. He had to determine quickly how to best meet his responsibilities to the air force and to the greater good. Something would have to be sacrificed. He decided to limit his investigation to the Mannor sighting of the twentieth and the Hillsdale College sighting of the twenty-first. Surprisingly, he dismissed the multiple sightings of the previous week, despite the fact that all the witnesses were law enforcement officers.

His faith in the local constabulary was shaken early on, however, when he found himself taking part in a frenzied pursuit of yet another light in the sky:

> The entire region was gripped with near-hysteria. One night at midnight I found myself in a police car racing toward a reported sighting. We had radio contact with other squad cars in the area. "I see it" from one car, "there it is" from another, "it's east of the river near Dexter" from a third. Occasionally even I thought I glimpsed "it."
>
> Finally several squad cars met at an intersection. Men spilled out and pointed excitedly at the sky. "See—there it is! It's moving!"
>
> But it wasn't moving. "It" was the star Arcturus, undeniably identified by its position in relation to the handle of the Big Dipper. A sobering demonstration for me.[26]

Equally sobering was the "photographic proof" offered by Deputy Fitzpatrick. "That photograph is another thing that set me off," Hynek said; "I took one look at that photograph all the papers had published, and it was obviously the Moon and Venus rising."[27]

Over the next three days Hynek interviewed twenty-eight people at length. His schedule included a side trip to the University of Michigan at Ann Arbor to interview eleven astronomy department faculty members (who wished to remain anonymous); a long, soggy trek through the swamp behind Frank Mannor's home, retracing the footsteps of Deputies McFadden and Fitzpatrick on the night of the twentieth; countless impromptu and unwelcome interviews with pushy reporters demanding explanations Hynek was not ready to give; and, lest we forget, one dizzying midnight ride in a police car in hot pursuit of that old troublemaker Arcturus. It would have been a trying three days for anyone.

When Hynek trudged through the swamp with Deputy Fitzpatrick, he learned that the deputy never actually witnessed a solid object. He had seen only a faint glow that rose and fell, but Fitzpatrick never saw it rise to treetop level as the Mannors had reported.

The inconsistencies mounted. Deputy Fitzpatrick claimed that the glow disappeared as he drew close and pointed his flashlight at it. He never saw the glow again, but the Mannors, safely back at the house, reported seeing it reappear instantaneously across the swamp.

Things started looking up when Hynek arrived in Hillsdale on Thursday. On paper, Bud Van Horn had all the observational and descriptive skills that the law enforcement officers in Dexter had been lacking. Not only was he the county civil defense agent, entrusted with the preparation of the citizenry in the event of a nuclear attack, but he was also a licensed pilot and an air force veteran.

Hynek met Van Horn at his home and found him "a very knowledgeable and effective person," as well as "generally articulate." After a lengthy interview, Hynek accompanied Van Horn to Hillsdale College, where Kelly Hearn and the dorm residents were waiting in anticipation. Hearn had decided to limit Hynek's access to only two of the eighty-seven women who had seen the lights in the arboretum, and her fervent protectiveness would have significant repercussions later in the investigation.

She led the men to room 204, where Sara Robechek and Jo Wilson recounted their experience from three nights earlier. "Both girls told me that attention was first called by a casual glance out of the window which lead [*sic*] to the remark by one of them 'there's no house down there, is there?' or words to that effect," Hynek said in his Blue Book report. "A fairly bright red light and two yellowish lights, resembling Christmas tree lights, according to the girls, were close to the ground in the marsh area, where, they obviously knew there was no house."

One detail in particular stood out to Hynek: all four witnesses stated that the lights in the arboretum were visible only when the lights in the dorm rooms were off.

Some of the other witnesses in the dorm had reported seeing a bright light sweep past their windows just before the main sighting, but none had seen a tangible object. "I could get no consistent story relative to this, in contrast to the fairly consistent description to the lights in the swamp," Hynek wrote.

As they looked out the window together, Van Horn described to Hynek how the lights periodically rose above the trees to a height of about 150 feet, growing brighter and then dimmer as they rose, then settled back down to ground level. Again, he characterized the lights as purposefully avoiding the airport beacon.

Then Van Horn made an especially interesting comment: "[He] stated that his first impression of the lights were that they were marsh lights, otherwise known as fox lights," Hynek reported.

Marsh lights—or fox lights, will-o'-the-wisps, jack-o'-lanterns, or orbs—can appear when methane and other gases formed by decaying organic matter in swampy areas rise in the air and oxidize, causing a faint floating glow and occasionally giving off a "popping" sound.

But marsh lights don't just glow and float and pop; like the

UFO phenomenon itself, they also appear to retreat when one approaches them. Because of this uncanny quality, marsh lights have been described in folk tales for centuries as possessing intelligence, even life. But life with a particularly mischievous bent: many a traveler on a dark night has found him- or herself hopelessly lost in a swamp, led astray by the glow of a will-o'-the-wisp that beckons but always remains out of reach. Find a fairy, and you may have found a marsh light.

By Van Horn's own account, he, too, was nearly led astray. He had been ready to explain away the apparition in the arboretum as marsh lights when the glow suddenly grew in intensity and started to rise for the first time. It was then that he thought he saw a "convex-shaped" solid mass between the lights and abandoned the marsh light theory.

"In Hillsdale, Mr. Van Horn was clearly the best observer as far as technical background was concerned, yet his knowledge of physics is not good . . ." Hynek reported. "Van Horn also struck me as a man who becomes enamoured with an idea and tends to lose objectivity in favor of the idea."

"I'm still gathering the facts," he later told a UPI reporter. Hynek "refused to speculate" on the nature of the Dexter-Hillsdale lights, leading the reporter to claim that his sleuthing was being "carried on behind Air Force secrecy."[28]

Although he told the reporter that these sightings were among the more consistent UFO reports he had investigated, Hynek's growing dissatisfaction with the inconsistency of the witnesses was beginning to show: "'It's like reports from people who saw a fire,' the astrophysicist said. 'You get as many different facts as you get people who saw the fire. So far, all I've been able to come up with is reports of a variety of lights.'"[29]

The lack of credible photographic evidence became particularly troublesome to Hynek once it became clear that the witnesses at Hillsdale College had had their reported UFO in sight for several hours. And why had no one ventured out to the arboretum to get a closer look at the lights? The second-floor dorm room afforded them a clear view of the phenomenon, but the lights remained

1,500 feet from the building for several hours. It would have taken only a few minutes to set out on foot to get a better view.

"I am responsible for all those young women," Hearn said, "and I wasn't about to let them go out there."[30] Nor was she willing to leave the girls to go out herself.

As for Van Horn, "I'd rather be a live coward than a dead hero," he later told a CBS interviewer, explaining that for all he knew the vehicle was emitting an electric charge that could have killed him had he gotten too close.

HYNEK'S INITIAL SUSPICION—that the Dexter-Hillsdale sightings did not appear to be as convincing as many other UFO cases he had in his files—was proving to be prophetic. Although there was a consistent theme among the Michigan sightings, the mass of details that emerged in interviews simply refused to gel.

"I was forced, therefore, to take into account, and regard as facts, only such things which had been reported fairly consistently by several witnesses," he said in his Blue Book report. He continued:

> I therefore decided to look at the common pattern which seemed to be developing in the Dexter and Hillsdale cases. These had many interesting points in common; first, the lights in both cases were associated with swamp areas . . . In both cases a distinctive reddish yellow, and green lights were reported, plus a general yellow glow . . . The similarity between the Dexter lights and the Hillsdale lights was striking. Not only as to color and intensity, but motion as well. The motion was described in both cases as smooth and slow, with a tendency to disappear rather suddenly and reappear elsewhere.

Those were the facts, as Hynek saw them, and they were very small in number. Furthermore, he reported, "The doubt cast upon certain witnesses by other witnesses did not inspire in me a feeling of confidence as to the quality and reliability of the data I was obliged to use from which to draw my conclusions."

And he was obliged to draw conclusions. "[Hynek] did not

want to issue explanations based on errors or probabilities or just skepticism—that was not an explanation," Powers said. "Either you had to say, 'Here is the actual fact behind it, with people who can vouch for it,' or 'I don't know.' 'I don't know' is the only alternative. That's what an *unidentified* flying object is!"

But "I don't know" was not going to wash in March 1966, in Michigan. And so, on the morning of the twenty-fourth—less than two days after Hynek had arrived on the scene, and still several hours *before* he was to interview the Hillsdale College witnesses—Quintanilla made a bold statement to the press that Hynek's investigation was "beginning to shape up. We hope to have a reasonable explanation sometime tomorrow."[31]

The stage was set. "Hector Quintanilla's policy of a rapid public disposition of cases was a public relations accident waiting to happen," said Swords and Powell, "and in late March 1966, it did."[32]

"In the midst of this confusion, I got a message from the Air Force: There would be a press conference, and I would issue a statement about the cause of the sightings," Hynek said. "It did me no good to protest, to say that as yet I had no real idea what had caused the reported sightings in the swamps. I was to have a press conference, ready or not."[33]

Major Quintanilla recalled the sequence of events quite differently. In his unpublished book manuscript, Quintanilla claimed he was in no particular hurry to wrap up the controversial case.

"[Hynek] asked me if he could have a news conference and I said no," Quintanilla wrote. "This was setting a precedent and I didn't like it."[34] What's troubling about this statement is that Quintanilla claimed that Hynek's request came "early in the morning" of March 23. This was Hynek's first full day in Michigan, before he had started his investigation. It seems absurd to suggest that Hynek would request a press conference before he had interviewed a single witness, but then, Quintanilla is fuzzy about dates throughout his manuscript. He claims, for instance, that the Frank Mannor sighting took place on March 17, when in fact it occurred on the twentieth.

Inaccurate dating aside, Quintanilla went on to write that

Hynek continued to call him at the Blue Book offices in Dayton requesting that Selfridge AFB set up a press conference. "The next day Hynek called again, and informed me that he had a possible solution to Frank Mannor's sighting and I asked him for the details . . ." Quintanilla wrote. "He told me that the solution was 'Swamp Gas.' I told him to check this out with his colleagues at the university and let me know their reaction. In the meantime, I would check it out with the chemists and botanists on the base.

"He also wanted me to arrange for a press conference from the Information Office at Selfridge AFB. I was against this from the beginning, but he was insistent and I told him I'd check it out with the Pentagon."[35]

In Quintanilla's account, he persevered, despite his professed misgivings, and persuaded his superiors to grant Hynek's request for a press conference, to be held that Friday, March 25.

Both Hynek's and Quintanilla's reluctance to take credit for what happened next is completely understandable, for it marked the end of the air force's credibility and the end of Hynek's innocence.

CHAPTER 13

WILL-O'-THE-WISP, PART TWO

THE PRESS CONFERENCE ITSELF was a high point for thousands around the country, and the world, who had long hoped for the UFO phenomenon to get this much attention and to be taken so seriously.

It was also, in Hynek's words, the absolute low point in his association with UFOs. "It was a circus," said Hynek, blaming the confusion on the public relations staff at Selfridge Air Force Base, whom he felt had bungled the entire event. "The TV cameramen wanted me in one spot, the newspapermen wanted me in another, and for a while both groups were actually tugging at me. Everyone was clamoring for a single, spectacular explanation of the sightings. They wanted little green men."[1]

There was, from the outset, a cognitive chasm between the speaker and his audience. The audience expected—needed—something sensational; the speaker gave only what would fit the facts. And what fit the facts was "swamp gas."

Swamp gas. He could not have uttered two more unexpected words.

"It would seem to me that the association of the sightings with swamps, in these particular cases, is more than coincidence," Hynek explained to the gathered reporters. "No group of witnesses observed any craft coming to or going away from the swamps. The glow was localized there."

He explained further that witnesses at both Dexter and Hillsdale described the exact same effects: the way the lights brightened and dimmed smoothly and slowly; the way they could be seen only in the complete dark, then went out when other lights appeared. "It appears to me that all the major conditions for the appearance of swamp lights were satisfied," Hynek reported.

"I emphasize in conclusion that I cannot prove in a Court of Law that this is the full explanation of these sightings," he said. "It appears very likely, however, that . . ."

The rest of Hynek's conclusion wafted off into the rafters of the Detroit Press Club. No one was listening anymore; the reporters were all filing their stories. They were all announcing to the world what the U.S. Air Force's number one UFO expert had just declared: that more than one hundred people in Michigan—policemen, college students, and civil defense authorities among them—had been duped by swamp gas.

It didn't help that a reporter shoved a photo of George Adamski's "spaceship" into Hynek's hand, and that image was caught on film and sent out over the wires. It looked to all the world as though Hynek was equating the Michigan witnesses with Contactees like Adamski.

The Michigan witnesses felt they had just been betrayed and humiliated by a snooty academic, and they were furious.

Dorm housemother Kelly Hearn said in another typewritten statement to the press:

> In the matters of astronomy, UFO's, and celestial calculations, I am a complete novice. Hence for me to take any issue whatsoever with Dr. Hynek's findings would be presumptuous, inappropriate, and ill-concidered [sic] on my part.
>
> I can say no more than this:
>
> 1. If the phenomenon were swamp gas as Dr. Hynek stated, then it behooves me to spend more time studying swamp gas and less time watching what I took to be a UFO.

    2. In my mind there exists that famous American prerogative—namely, a reasonable doubt.

"It was my considerate [*sic*] opinion that Dr. Hynek had his mind made up as to what his findings would be before he ever reached the City of Hillsdale," said Bud Van Horn. "I also observed that his main line of questioning was relative only to that which would fit the Marsh Gas theory."

"I'm just a simple fellow," said Frank Mannor. "But I seen what I seen and nobody's going to tell me different. That wasn't no old foxfire or hullabillusion. It was an object."[2]

Was the criticism of Hynek valid? Certainly not all of it, as a look at Bud Van Horn's statement to the press reveals.

Van Horn Criticism: "Although there was nothing to my knowledge from the information given Dr. Hynek that would fit the Marsh Gas theory, he irregardless [*sic*] found it fit to state that Marsh Gas is what we were observing." The problem with Van Horn's criticism is that Van Horn himself told Hynek that he thought the lights were marsh gas when he first saw them. He literally planted the idea in Hynek's head.

Van Horn Criticism: "Dr. Hynek was not or at least didn't to me display any intent interest in the type of movement of the Vehicle that we were observing nor did he volunteer any thought or explanation of the observation the girls had made of the object descending from the sky which prompted them to call the Civil Defense office." The problem was Kelly Hearn had limited Hynek to interviewing only Sara Robechek and Jo Wilson, both of whom testified that they had first seen the lights at or near ground level in the arboretum and at first took them to be a house. Therefore, there was no description of "the object descending from the sky" for Hynek to think about or explain.

Van Horn Criticism: "At no time did Dr. Hynek step a foot into the area where the Vehicle was located that the girls and myself had been observing." The problem was neither had he.

Yet that didn't stop Van Horn from identifying the lights as a "Vehicle." How did he know? He had already stated that when he drove past the arboretum on his way to the dorm, he had been searching for the lights but hadn't seen a thing. Furthermore, Hearn's scolding Hynek about "reasonable doubt" completely overlooked the professor's statement to the press that he could not prove the swamp gas theory in a court of law.

But by making the decision to consider only the Mannor and Hillsdale College sightings and then to consider only those facts that could be corroborated by more than one witness, Hynek put himself in a position where he was guaranteed to make many, if not all, of the witnesses feel cheated and belittled. He also knew that his statement in Detroit would offend and anger everyone he had ever recruited to the UFO cause.

"Hynek has just called me: he felt the need to explain to me his Detroit statement, knowing that I was seriously disappointed, like everybody else," wrote Jacques Vallee. Vallee insisted to Hynek that with the testimony of eighty-seven students of Hillsdale College behind him, he could have made a strong case at the press conference that "there were clear patterns" that demonstrated the need for a serious study.[3]

But he had the testimony of only two of the students, not eighty-seven, and the patterns at Dexter and Hillsdale weren't so clear to Hynek, something that Vallee seemed to have a difficult time appreciating.

"Too many of the stories the next day not only said that swamp gas was definitely the cause of the Michigan lights but implied that it was the cause of other UFO sightings as well," Hynek lamented. "I got out of town as quickly and as quietly as I could."[4]

Had Hynek been more concerned about his newly tarnished public image, he could easily have deflected a great deal of the criticism and ridicule. All he had to do was tell the press that Bud Van Horn had originally suggested the swamp gas solution, and all the heat would have been on Van Horn.

But Van Horn wasn't his only source. From Hynek's lengthy case notes and Blue Book report, it becomes evident that as many as a half dozen people mentioned the swamp gas theory to Hynek as a likely explanation for the Dexter-Hillsdale sightings during his three days in Michigan. The list of swamp gas sources includes University of Michigan astronomy, chemistry, and botany professors; some unnamed "Michigan scientists"; an anonymous military source; and Hynek's Blue Book colleague Task Sergeant Moody.

Strangest of all was a small item that appeared several hours before the press conference, in the Friday, March 25, edition of the *Detroit Free Press*. The article said that the air force "expects to come up with a reasonable, logical answer sometime Friday."[5] Then it dropped a bomb, saying that Dr. Hynek had admitted to finding no evidence to suggest that the Dexter-Hillsdale lights were the result of "extraterrestrial intelligence." The unnamed reporter predicted that the air force's "official" conclusions at the press conference would not satisfy the many witnesses in the area, and then cited one Alfred Dickens, a maintenance man for the York County Gas Company of York County, Pennsylvania, who "snorts at saucers and says the phenomena are merely balls of 'damp gas' seeping from swamps and marshes."[6]

So it seems that the swamp gas "seed" could also have been planted by Major Quintanilla, masquerading as "Alfred Dickens." That the article—perhaps penned by Quintanilla himself—was a plant by the air force to influence the outcome of that evening's press conference seems clear. The question is whether the air force planted the article that morning to prime the media or to box Hynek in and pressure him into making the swamp gas statement.

IN THE WAKE OF HYNEK'S PRESS CONFERENCE, U.S. congressman and House Minority Leader Gerald R. Ford and Congressman Weston Vivian of Ann Arbor were urged by their angry constituents to call for a congressional hearing into the air force's handling of the Dexter-Hillsdale sightings. The congressmen agreed, feeling that the air force had failed to give the American public a satisfactory explanation for the sightings and that the Congress should hold the air force accountable.

Their demand met with resounding and surprising success, and the House Armed Services Committee scheduled its first open hearing on the UFO phenomenon a mere two weeks after the sightings. For better or worse, the genie was out of the bottle.

The air force brass, so used to being unquestioned on matters of national security, were caught short. According to Vallee, Hynek was delighted: "'You ought to see the Air Force people,' he said with glee, 'they're running around like chickens with their heads cut off!'"[7]

When the hearing began, only three witnesses were called: Secretary of the Air Force Harold D. Brown, Dr. Hynek, and Major Quintanilla. In his prepared statement, Brown acknowledged that there were 648 "unexplained" reports in the Blue Book files, but he insisted that there was no evidence to prove that UFOs were an extraterrestrial phenomenon or posed a threat to the United States.

Still smarting from his treatment in the press over the previous weeks, Hynek "warned that complete adherence to the policy that all UFO reports had conventional explanations 'may turn out to be a roadblock in the pursuit of research endeavors,'" and "called for a civilian panel of scientists to examine the UFO program critically and to determine if a major problem actually existed."[8]

In his book, Quintanilla accused Hynek of a major breach of trust, claiming he did not have his statement cleared by General Thomas Corbin of the Chief Legislative Liaison Office at the Pentagon before the hearing. "He had told Corbin that he was not going to make a statement and then he pulled out a five page neatly typed statement from his briefcase," Quintanilla wrote. "As far as I was concerned, he had deliberately and with premeditated motives lied to General Corbin. I had been losing confidence in Hynek for some time and after the hearing he never regained my original confidence."[9]

When his own turn came, Quintanilla offered deceptive responses to the questions that came his way. For example, when Pennsylvania congressman Richard Schweiker asked Quintanilla whether any unexplained objects had been sighted on radar, the major said, "We have no radar cases that are unexplained."[10]

Hynek was outraged. Some of the most compelling unexplained UFO sighting reports in Project Blue Book's files involved radar detection, and Quintanilla was telling Congress that no such thing existed. "I almost felt like getting up and saying, 'You lie!'" Hynek said, "but this was Congress and I didn't."[11]

Quintanilla alone had the power to declare cases "explained" or "unexplained," so in one sense what he told the congressman was true: there were no unexplained radar cases because he declared them all "explained."

Hynek's call for a civilian panel to study the phenomenon was heard, however, and acted upon: "The committee realized it couldn't deal in one day with a problem that had bamboozled the Air Force for almost 19 years and 'suggested' that the Pentagon set up a civilian inquiry into flying saucer phenomena."[12] Secretary Brown allowed that he had already been considering this course of action, and when the day ended, a plan was in place to recruit a scientific panel that would at last treat UFOs as serious business.

DESPITE HIS BEST EFFORTS, Bud Van Horn did not prove Hynek wrong about the swamp gas case. In fact, it took little time for the doctor to ride out the backlash that had been unleashed against him by Van Horn and others in the wake of the Detroit press conference. Hynek's correspondence files are brimming with letters mailed to him in the spring, summer, and fall of 1966, many asking for more information about UFOs, many more asking for his expert analysis of the writers' unique UFO experiences. But now, more and more, letters were being addressed to Hynek at Dearborn Observatory, not Project Blue Book.

In the aftermath of the Dexter-Hillsdale sightings, the failure of the air force's nineteen-year UFO public relations program was as complete as it was unlikely. After all, one could imagine the demise of Project Blue Book coming in the wake of aliens from space landing their flying saucer on the White House lawn, but the fact that Blue Book's credibility was finally shattered by rotting vegetation was the ultimate irony.

At the same time, the success of Hynek's personal PR game was just as complete. He had emerged from Michigan as the world's preeminent UFO expert.

"Every day there are more cartoons about Hynek, ridiculing his 'marsh gas' explanation," Vallee recorded in his journal ten days after the hearing. "Oddly enough, this has the effect of placing him increasingly in demand as a lecturer. People want to see and touch him, and he enjoys every minute of this newfound celebrity."[13]

Even Quintanilla was in awe. "The publicity that this sighting received was unbelievable," he wrote. "Hynek became an instant celebrity and the sightings started pouring in. We had a total of 1,112 sightings in 1966 and that total has never been equaled since."[14]

Still, there were always regrets about Michigan. "I think the swamp gas thing was sort of a turning point," Powers reflected. "He got ridiculed a lot, and that was hard. Because he was a scientist, and to hear yourself described as sort of a stiff-necked, pooh-poohing debunker was not what he wanted to be. He didn't think of himself that way."

THE O'BRIEN COMMITTEE HAD RECOMMENDED that a group of universities be selected by the air force to study UFOs, with one university coordinating the work. Less than two months later, the House Armed Services Committee recommended the same. But when Secretary Brown assigned the Air Force Office of Scientific Research to set about recruiting universities to participate, it found no takers. "No" from MIT. "No" from Harvard. "No" from the University of California. Twenty-five universities were approached, and they all said no. "UFOs were leprosy to academia," said Swords and Powell.[15]

Hynek's own university was not interested. Northwestern's administrators were already uneasy about their employee's growing visibility as a UFO expert, but there was another complication as well. The Corralitos Observatory had just been dedicated a few months earlier and the new Lindheimer Astrophysical Research Center was now rising on the shore of Lake Michigan. Northwestern's PR office wanted the press's attention—and Hynek's—to be on the new facilities.

Even if the timing hadn't been so inconvenient, Hynek was far too close to the subject to be part of what was intended to be an unbiased study. "The reason I wasn't on the committee is that the Air Force wanted a 'fresh start,'" Hynek explained.[16]

He was by no means sidelined, however. As spring turned into summer and the air force struggled to recruit even one university, Hynek maintained a heavy schedule of lectures, interviews, and public appearances. Vallee and Powers worried about his health and energy level, and they pressed him to give up the limelight— even resign from Project Blue Book—so that he could concentrate on their own research activities.

"I know that you regard these lectures as a responsibility to inform the public," Vallee told him at the beginning of May. "But the events that are taking place demand that you stay here, on the bridge, to steer this ship."[17]

Hynek couldn't be dissuaded, however, even when his public appearances failed to further his cause. Later that month, for example, he appeared on the CBS television news special *UFO: Friend, Foe or Fantasy,* a sixty-minute exercise in reassurance designed to ease the country's fears in the wake of the Dexter-Hillsdale sightings and the book *Incident at Exeter.*

"You might call me a study in puzzlement," Hynek said in his interview segment, responding to repeated questions about whether spaceships were visiting Earth. "But among the tremendous noise, or static, or crud, or whatever you want to call it—the tremendous number of unreliable reports that are easily explained—there is this residue of most interesting cases that intrigue me, the same way a good mystery story intrigues me, and I'd like to get the solution.

"I don't think it is space people," he quipped, "although I would be delighted if it turned out that way, because as an astronomer I would think it would make astronomy even more interesting than it is." His last words on the telecast could have been directed at Bud Van Horn: he said that he was fine with anyone *claiming* that UFOs are alien spaceships, but that the burden of proof rests with them.

In the final segment of the show, astronomers Carl Sagan (member of the recent O'Brien Committee) and Thornton Page (part of the Robertson Panel of 1952) insisted that astronomy was quite interesting enough on its own without having to consider space people. While Page smirked, Sagan reduced the belief in "the UFO myth" to a demeaning need for humans to believe in benevolent, omnipotent beings in "long white robes" that will save us from ourselves.

Although Page remained largely silent in the interview segment, he remained culpable, as he had actually helped develop the show for reporter Walter Cronkite. While acknowledging the importance of Hynek's UFO work, Page maintained, "My own inclination is that [UFOs] are natural."[18]

Cronkite seemed to agree with Sagan and Page, closing the program with this wisdom: "We might remember, too, that while fantasy improves science fiction, science is more often served by facts."

Vallee was incensed. "[Hynek] was the only sincere man in the whole bunch, the only scientist genuinely searching for the truth . . ." he wrote after watching the telecast. "But the editing of the documentary made dialog and rebuttal impossible, as usual."[19]

Hynek, too, was growing frustrated. The more his words were edited, the more his position on the UFO phenomenon was misunderstood, and the more his position was misunderstood, the more anger he inspired. And three months after the swamp gas case, that anger showed up in his office, yelling, scolding, and banging its fist on Hynek's desk.

Dr. James McDonald, professor of atmospheric physics at the University of Arizona, saw a UFO in 1954 while driving with friends in the desert. When he was displeased with the way Project Blue Book handled his report, and neither APRO nor NICAP seemed up to the task of conducting a proper investigation, McDonald used his connections to gain access to the Blue Book files. What he found at Wright-Patterson on his June visit shocked him.

There, in a copy of the 1952 Robertson Panel report—which had been seen by very few—McDonald found the panel's recommendation that the U.S. Air Force should use Project Blue Book to debunk UFO reports and thus dampen public curiosity about the subject.

Two days later, when McDonald arrived in Evanston to pay a call on Hynek, he was in a rage. McDonald was disturbed by "all the good UFO cases that Hynek had dismissed with absurd explanations"[20] and challenged his colleague to explain himself. "[McDonald] thought I was so completely dead wrong and so obtuse!" Hynek said of the encounter. "He . . . pounded on the table and he said, 'Allen, how could you have sat on this data for eighteen years and not let us know about it?' And I said, 'Well, damn it, what data? Large parts of the data are sheer nonsense!'"[21]

Despite Hynek's chagrin, he was on one level absolutely thrilled by McDonald's visit: here was another establishment scientist demanding that UFOs be studied properly!

"You have no idea what a relief it was, like taking off a pair of very tight shoes, to have a sympathetic listener," Hynek said. "There was not one of my colleagues interested in it, not even on my own staff. All those people were absolutely convinced I was a nut. So, the very fact that Jim McDonald existed was a relief, even though he thought I was wrong and should have gone and pounded on the doors of the generals."[22]

Hynek made the case that if he were to pound on those doors the air force would simply replace him with someone more compliant, and he would forever lose access to the riches in the Blue Book files. "My temperament is to play the waiting game," Hynek admitted. "I always have, maybe it's my Czech background."[23]

That summer, however, perhaps spurred on by McDonald's anger, Hynek declared an end to the waiting game. On August 1, less than two months after McDonald's visit, he officially broke cover and submitted a lengthy letter about UFOs and Project Blue Book to the editors of *Science,* the academic journal of the American Association for the Advancement of Science (AAAS). Apparently cured of his Czech background, Hynek made sure his first assault on Project Blue Book would register at the heart of the scientific establishment.

The letter appeared in the October 1966 issue of the journal, where, even in its abbreviated form, it took up an entire page. Hynek had a lot to get off his chest.

"I feel under some obligation to report to my scientific colleagues, who could not be expected to keep up with so seemingly bizarre a field, the gist of my experience in 'monitoring the noise level' over the years . . ." he began. "In doing so, I feel somewhat like a traveler to exotic lands and faraway places, who discharges his obligation to those who stayed at home by telling them of the strange ways of the natives."[24]

Dr. Hynek looking bemused, perhaps after digging through the latest batch of UFO reports from Project Blue Book. Note the example of what his son Paul referred to as his father's "astro-beatnik" neckwear. *(Photo by Tom Brunk. Courtesy of Northwestern University.)*

After a description of his long tenure as a UFO investigator, Hynek recounted his frustrated attempts to nudge the air force into taking a more scientific approach to the phenomenon. He then declared that it was time at last to clear up popular misconceptions about UFOs, and he addressed seven that he felt were the most egregious:

1. *Only UFO "buffs" report UFO's.* The exact opposite is much nearer the truth . . . It has been my experience that quite generally the truly puzzling reports come from people who have not given much or any thought to UFO's.

2. *UFO's are reported by unreliable, unstable, and uneducated people.* This is, of course, true. But UFO's are reported in even greater numbers by reliable, stable, and educated people . . .

3. *UFO's are never reported by scientifically trained people.* This is unequivocally false. Some of the very best, most coherent reports have come from scientifically trained people. It is true that scientists are reluctant to make a public report . . .

4. *UFO's are never seen at close range and are always reported vaguely.* When we speak of the body of puzzling reports, we exclude all those which fit the above description . . .

5. *The Air Force has no evidence that UFO's are extraterrestrial or represent advanced technology of any kind.* This is a true statement but is widely interpreted to mean that there is evidence against the two hypotheses. As long as there are "unidentifieds," the question must obviously remain open . . .

6. *UFO reports are generated by publicity.* One cannot deny that there is a positive feedback . . . when sightings are widely publicized, but it is unwarranted to assert that this is the sole cause of high incidence of UFO reports.

7. *UFO's have never been sighted on radar or photographed by meteor or satellite tracking cameras.* This statement is not

equivalent to saying that radar, meteor cameras, and satellite tracking stations have not picked up "oddities" on their
scopes or films that have remained unidentified.[25]

His argument was comprehensive; his points were sound. How
could any thinking person not be persuaded to at least consider
the data?

"I cannot dismiss the UFO phenomenon with a shrug," he
concluded, adding, "I have begun to feel that there is a tendency
in 20th-century science to forget that there will be a 21st-century
science, and indeed, a 30th-century science, from which vantage
points our knowledge of the universe may appear quite different.
We suffer, perhaps, from temporal provincialism, a form of arro-
gance that has always irritated posterity."[26]

Six months and one day from the date that the House Armed
Services Committee directed Secretary Brown to recruit a group
of universities to study the UFO phenomenon, the first school
signed on. Relieved to have one contract in place, the air force
quietly abandoned its attempts to attract any other institutions.

The University of Colorado stepped up to the plate after being
led to believe that it was one of the air force's first choices for the
study. Add to that flattery a generous $300,000 grant and a waiver
from the standard matching funds requirement, and the university could hardly say no.

So, at about the same time that Hynek's letter was going to print
in *Science* magazine, the world learned that Dr. Edward U. Condon, respectable physics professor at the University of Colorado
and former director of the National Bureau of Standards, was to
chair a committee that would settle the UFO issue once and
for all.

Hynek's joy at the Condon announcement was palpable. The
spotlight was off him for the moment, but it didn't matter because
something was finally being done.

"I cannot help but feel a small sense of personal triumph and
vindication," he wrote in the *Saturday Evening Post*. "The night

the appointment was announced, my wife and I went out and had a few drinks to celebrate . . .

"Now after a delay of 18 years, the Air Force and American science are about to try for the first time, really, to discover what, if anything we can believe about 'flying saucers.'"[27]

The wild ricochet of the swamp gas statement was a wonder to behold.

CHAPTER 14

MR. UFO

IN THE YEARS SINCE their strange nighttime encounter along the highway in New Hampshire, Betty and Barney Hill had been compulsively returning to Indian Head to retrace their movements from that night. Somehow, somewhere, the Hills had lost two hours of their lives on that trip, and they were determined to find out why. Every week became a slow, tortuous countdown to the weekend, when they could once again toss their dog, Delsey, and their binoculars into the car and duplicate their drive through the mountains. After years of nightmares and anxiety, Barney and Betty now suspected that there was a side road somewhere along Route 3, an unmarked road onto which they were diverted that night, but no matter how many times they repeated their drive they could never find any trace of the turnoff.

The nagging doubts took their toll, and Barney's health began to deteriorate: his ulcers flared up, he suffered from high blood pressure, and an odd circle of warts had appeared on his groin. The incident was affecting Betty as well; she was afflicted by terrible dreams of being abducted by strange men in the road.

In early 1963 the Hills were referred to Boston psychiatrist Dr. Benjamin Simon, a well-known expert in hypnotherapy. Based on his initial conversation with the Hills, Dr. Simon decided to treat both Barney and Betty. They seemed to present a unique case of shared amnesia, perhaps shared hallucination, and Dr. Simon was intrigued.

At first, the Hills outright refused to remember what had happened that night after they raced away from Indian Head, even under hypnotic suggestion. Whatever transpired during the two missing hours was too terrible for them to face, and they resisted with all their might; in the transcripts of the hypnotic sessions, Barney would beg Dr. Simon, "Please can't I wake up?"[1] so he wouldn't have to remember. In one early moment Barney described the intensity of his terror by equating himself with a rabbit he remembered from his youth: "I was hunting for rabbits in Virginia. And this cute little bunny went into a bush that was not very big . . . And the poor little bunny thought he was safe. And it tickled me, because he was just hiding behind a little stalk, which meant security to him—when I pounced on him, and threw my hat on him, and captured the poor little bunny who thought he was safe."[2]

Eventually, resistance broke, and the story of the phantom side road and the missing two hours took on a solid, disturbing form. Week by week, detail by detail, Betty and Barney recalled under hypnotic trance—separately, without any knowledge of what the other had said—what had happened that night in 1961:

A few miles beyond Indian Head, after hearing strange beeps from the back of their car, Barney and Betty saw red lights on the road ahead and thought there must be an accident blocking the highway. They were somehow diverted off Route 3, and Betty remembered Barney making a sharp left-hand turn. Once on the side road, they were flagged down. They slowed to a stop and the car died. A group of six men in the road approached the car. They split up into two groups and "assisted" Barney and Betty out of the car. A brilliant orange light like the moon was visible through the trees.

Barney was carried from the car, his feet dragging on the ground. He had been told to keep his eyes shut and so he did. "I thought how funny," Barney recalled, conjuring up the memory of the bunny. "If I keep real quiet and real still, I won't be harmed."[3]

Betty was escorted by several of the men into the forest and toward the glowing moon. She was aware that Barney was being carried along by two men behind her, but he did not respond to her and she thought he must be asleep. One of the men assured Betty that she and Barney would not be harmed; the men merely wanted to conduct some tests on them, after which they would be returned to their car. "You'll be on your way back home in no time," the man told Betty.[4]

About the men: They were short in stature; they had large, dark, slanted, teardrop-shaped eyes that curved around the sides of their heads; they had grayish skin; and they wore dark uniforms. They were able to communicate in a way that did not rely on audible speech, although they were capable of speech and spoke English with an unusual accent. And they looked very little like the "light being" from *The Outer Limits*.

Arriving at the orange moon, the men walked Betty and Barney up a ramp and through a door. Betty was directed down a corridor—again a left-hand turn—and taken to a room. She was upset to see Barney being taken to another room, but was told by the man she came to refer to as the "leader" that there was not enough space to examine them both in the same room.

Without any further fuss, the examination began. First, the man she came to regard as the "doctor" took a skin scraping, then swabbed her ear—the left one. Then he took a few strands of Betty's hair and cut off a small piece of her fingernail. Despite her fear and confusion, Betty remained compliant, and curious, as the doctor had her take off her shoes and dress and lie back on a table. The doctor ran needles over Betty's head, neck, knees, ankles, stomach, and back. He then inserted a very long needle into Betty's navel, and the leader explained that it was a pregnancy test. When Betty cried out in pain, the leader placed his hand over her eyes and the pain disappeared. This gesture created a powerful bond of trust between Betty and the leader.

As the examination ended and the doctor left, Betty felt some anxiety about Barney, but the leader said Barney needed more tests and would "be right along in a minute."[5] After she put her clothes back on, Betty asked the leader to do her a favor, and so began one of the most disarmingly human conversations in all of UFO lore. Explaining that no one would ever believe her story, she asked the leader to give her proof that the abduction had really happened. The leader laughed and told Betty to pick out something on the ship. She chose a large book with writing inside that was arranged top to bottom and that looked to Betty somewhat like Japanese characters. Delighted with her gift, Betty then asked the leader from where he and his people came. After she admitted that she knew practically nothing about the universe, the leader showed her a star map that depicted the trade routes and exploration routes his people used. When Betty asked which of the circles on the map was his home star, the leader demurred. He said that since Betty didn't know which one was her own home there wasn't much point in showing her where he came from.

The doctor then returned and demanded to see Betty's teeth. She opened her mouth and he tried to tug her teeth loose, but they wouldn't budge. The puzzled doctor explained that Barney's teeth had come out, and he asked Betty why hers did not. Through her laughter, Betty tried to explain Barney's dentures to the men, but they could not understand.

After more talk in which Betty learned that the men also didn't seem to understand aging, the concept of a year, or what a vegetable was, Betty and Barney were released and returned to the car. Before she left, however, the leader made Betty return the book. The other men objected to him giving away something that would allow her to remember the incident, and she reluctantly surrendered her proof. She was, she told Dr. Simon, "furious."[6]

Barney's recollections were far less elaborate, but no less chilling, than those of his wife. Still channeling the bunny, he had remained quiet and kept his eyes closed as the men took him into a room and placed him on an exam table. One of the men, presumably Betty's "doctor," felt along Barney's spinal column with his fingertips, seemingly counting the vertebrae, then pressed hard at the base of Barney's spine. Then the man touched Barney's teeth, and a number of other men clustered around the left side of the table. Barney felt something covering his groin, then felt very light scratching along his left arm. After the examination his shoes were put back on his feet and he was led to the door. His feet scraped across a high doorjamb, as they had when he had entered, and then he was walking back down the ramp and toward the car. He got in the Chevy and found Delsey curled in a tight ball under the seat. He picked up the pistol, which he had left on the front seat, and thought, "That's interesting. What's this doing here?"[7] He placed it between the seat and the door.

A few moments later Betty came walking toward the car, grinning, and got in the passenger seat. They watched the glow from the woods grow brighter and brighter, and, as Barney said, "away it went."[8]

He started the car and drove back to Route 3. Barney had already forgotten everything.

Dr. Simon was in a fix. He could rule out lying and hallucination, but where did that leave him? He had not allowed Betty or Barney to consciously recall any of the details of their hypnotic sessions thus far, and the ramifications of that were immense: the Hills had both told virtually identical stories of an event so fantastic as to be completely outside the range of known human experience, and they had done so without any knowledge of the other's testimony. While unwilling to describe their experience as either illusion or reality, Dr. Simon was able to accept "the probability that the Hills had had an experience with an unusual aerial phenomenon, a sighting that stimulated an intense emotional experience for both of them."[9] He felt that the abduction experience, however, was "improbable."

BY EARLY 1966, four and a half years since the Hills had had their strange encounter, Barney and Betty were well known in UFO circles in New England. Indeed, they had found some solace and a sense of community in sharing their experience with others. Inevitably, word of their experience reached the popular press, and when a Boston newspaper printed a sensational but inaccurate account of their abduction, Barney and Betty became unwilling international curiosities.

Hynek, fresh off the swamp gas fiasco, was finding that his newfound fame acted as a mysterious attractor for high-profile cases like the Hills'. Because so much of the Hills' story depended on their hypnotic sessions with Dr. Simon, it perhaps comes as no surprise that Hynek's first encounter with the couple involved Dr. Simon and a bit of hypnosis.

Hynek initially put the Hills' case in the same category as the Contactees, although he was struck by the fact that their story "has no pseudoreligious, UFO cult overtones, no platitudinous cosmic messages of little content."[10] After John Fuller's bestselling authorized account of the Hills' experience, *The Interrupted Journey,* came out in the fall of 1966, Hynek was invited by Dr. Simon to dine with him, Fuller, and the Hills at Simon's home outside Boston. This was to be no ordinary dinner party, though, for Simon had also offered Hynek the opportunity to question the Hills while they were under hypnosis. "For the next hour and a half they went through their whole experience," Hynek recalled, "and at the time when Barney was supposed to be bodily dragged into the UFO, he cried out with such terror and such anguish, that I just couldn't *think* that he could have been faking that."[11]

Although he would eventually describe the Hill event as "perhaps the classic UFO abduction case,"[12] Hynek struggled with it even after his dinner with Simon, Fuller, and the Hills. There were, of course, numerous apparent inconsistencies in the story: If the men didn't understand the concept of a year, why did the leader use expressions like "in no time" or "in a minute"? Why would the men make the Hills forget the second encounter but let them remember the first? Why did they speak to Betty but use telepathy to communicate with Barney?

Hynek wasn't struggling with disbelief, however: the Hills' encounter was simply the first of its kind within his experience and therefore did not fit any identifiable pattern. Yes, he had in recent years been confronted with several "occupant" cases, but abduction cases were atypical to him; missing time was atypical to him;

amnesia and hypnosis were atypical to him. "When and if other cases of hypnotic revelation of close encounters become available for study (one recalls that the Hills waited several years before seeking treatment), we will be able to note whether they also form a pattern," he wrote.[13] Until that time, well, Hynek could at least be counted on to keep an open mind.

He hadn't counted on Marjorie Fish, however. A schoolteacher and Mensa member from Ohio, Fish had become fascinated by the Hill case, and when Betty's drawing of the alien star map appeared in *The Interrupted Journey,* Fish decided to give it the Rand McNally treatment. What she found after several years of research remains unexplained and controversial to this day: using hanging arrangements of beads and string, Fish was able to identify a starscape well within Earth's galactic neighborhood that matched the arrangement of stars in Betty's map. The Hills' alien abductors, Fish conjectured, could very well have come from the system of the double star Zeta Reticuli.

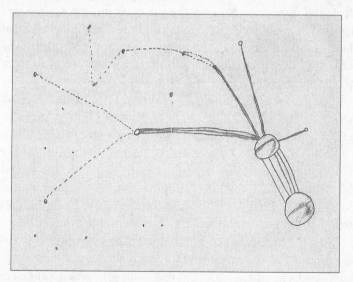

The star map allegedly shown to Betty Hill by one of her alien abductors, and reproduced here by Hill under hypnosis, was later found by researcher Marjorie Fish to correspond to stars in the neighborhood of Zeta Reticuli. *(Artwork by Betty Hill. Courtesy of Kathleen Marden.)*

"**I THINK WE'VE GOT** a real UFO on our hands."

That was the call of an air force intelligence officer before scrambling a fighter to get a visual on the blip that just appeared out of nowhere on radar. "UFO is picking up speed and climbing," the fighter pilot reported upon getting a radar fix. "I'm going in closer."

"I can see it now," he said a moment later. "Whatever this thing is, it's big."

Within moments the base lost contact with the plane, and it disintegrated. The pilot was married and had two young children.

If this reads like a replay of the Mantell crash of 1948, that's because it is. Only this time the drama was being played out in an episode of *Star Trek,* a topical new science fiction TV series that premiered with great fanfare within weeks of Hynek's 1966 meeting with the Hills. The series—the first American TV show to depict a multicultural, multi-gender, and multispecies space crew facing complicated dangers in space with thoughtfulness, tolerance, and reserve—would rewrite the book on how science was depicted in popular culture. From its humble beginning as one of the lowest-rated series on network television, *Star Trek* planted the seeds of modern geek culture and inspired a generation of children to pursue careers in science and work for NASA.

The first-season episode "Tomorrow Is Yesterday" traded openly on the Mantell case to tell its story of a 1960s American fighter pilot whose fate is altered when he pursues a massive UFO. Only on TV, the UFO is actually the starship *Enterprise,* and the pilot doesn't actually die. Instead, Captain Kirk, Mr. Spock, and the resourceful crew of the *Enterprise* "beam" the pilot aboard their ship an instant before his jet breaks apart, giving him an unauthorized peek at the future and thus upsetting the timeline.

How the *Enterprise* and its crew find themselves over U.S. airspace in the 1960s is explained by a "black star" and a "time slingshot," two very shaky concepts that somehow sound altogether logical and even inevitable when explained by Mr. Spock. The real point of interest here is that when *Star Trek* premiered in the 1966–67 TV season, creator Gene Roddenberry and writer Dorothy "D.C." Fontana recognized that UFO lore had a very strong hold on the public's imagination, and they were not afraid to build an entire episode around a UFO case from nearly twenty years earlier. Clearly, they knew the story would resonate with their viewers. It would seem "real."

It comes as a pleasurable surprise that Captain Kirk himself knows a bit about twentieth-century UFO politics. "If I remember my history correctly," he says upon realizing that the U.S. Air Force considers the *Enterprise* a UFO, "those were all dismissed as weather balloons and sundogs . . . explainable things. At least, *publicly.*"

Captain Kirk may have been referencing the government's new UFO study, the Colorado Project, which opened for business in the autumn of 1966 at the University of Colorado. The project was more commonly called the Condon Committee, in honor of its chair, physicist Edward U. Condon, who had assembled a dream team of U of C professors, including psychologists, physicists, astronomers, a meteorologist, a chemist, an electrical engineer, and a staff of impartial UFO investigators. The committee would, in effect, be taking over UFO case investigations from the air force for the duration of the study.

Condon himself got off to a shaky start with the American public by telling one reporter that he was a UFO agnostic and another that he wouldn't believe in flying saucers until he saw one himself and took it apart in a lab. To make matters worse, the project administrator, Assistant Dean Robert Low, came close to delegitimizing the whole enterprise when he told the press that the UFO study barely met the university's standards for acceptable research projects. While Condon issued retractions and corrections, the project started to take shape.

"It was obvious that . . . it would be necessary to bring prominent persons involved in the study of flying saucers to Boulder to brief the entire staff," wrote Dr. David Saunders, one of the psychologists on the committee. "Logically, the first man to interview seemed to be 'Mr. UFO,' Dr. J. Allen Hynek, of the Northwestern University Department of Astronomy."[14]

Mr. UFO was feeling exultant when he arrived in Colorado the night before he was to address the committee. He had insisted that the committee extend the invitation to present to Jacques Vallee, and Vallee had accompanied Hynek from Northwestern. Both men believed they had important contributions to make to the Colorado Project and were feeling deservedly self-congratulatory over their invitations to speak before the committee.

Driving through Denver in their rented sports car, the two scientists thought the UFO phenomenon was finally going to be accorded the respect they had long felt it deserved. Their appearance before the committee was, in Hynek's words, "a historic occasion,"[15] one for which he had been waiting nearly fifteen years. His sense of triumph and vindication, like fine champagne, made him feel giddy and loosened his lips.

"Did I ever tell you how I became interested in science?" Hynek asked his friend.[16]

"Wasn't your mother a schoolteacher?" Vallee responded, thinking he already knew all that Hynek had to reveal. "You told me she once gave you a book about astronomy that fascinated you."

"That's not what made me decide to take up science as a profession," Hynek confessed. "So many people get into science looking for power, or for a chance to make some big discovery that will put their name into history books . . . For me the challenge was to find out the very limitations of science, the places where it broke down, the phenomena it didn't explain."

The conversation then took an unexpected turn, with both men revealing their deepest spiritual beliefs. It was as though the two were speaking for the first time.

Hynek spoke unabashedly of his fascination with "esoteric subjects," admitting that in his youth he sought out the writings of the Rosicrucians and the hermetic philosophers. Now it was Vallee's turn to "take a deep breath," and he confessed that until recently he had sought illumination through the teachings of the Rosicrucian Order. The two weighed the merits of the various Rosicrucian teachers, and Hynek admitted to his fascination with the supersensible realm of Rudolf Steiner, whom he felt was "the deepest of the group."[17]

It is one thing to reveal an unusual, disreputable belief, and quite another to have it accepted and even shared. One can only imagine the relief that settled over their car.

Hynek thought awhile, looking out at the dark silhouettes of the Rocky Mountains. "I never cease to be fascinated by the limitations of our science, as great and amazing and powerful as it has become," he said. "Now we are about to see how it handles this phenomenon of UFOs that has become so familiar to you and to me. Yes, we shall see . . ."[18]

CHAPTER 15

SIGNAL IN THE NOISE

CONDON COMMITTEE MEMBER DAVID SAUNDERS knew only enough about Mr. UFO to know that he was seen as "an enigma to some."[1] It was easy to understand why; after all, Hynek had, since 1952, been calling for a serious scientific investigation of the UFO phenomenon, yet just a few months ago he had "affronted the residents of Washtenaw County, Michigan, with the famous swamp gas explanation."[2]

Observing that some had charged that "Hynek's position of the moment is determined by a careful check on the wind's direction at the Pentagon," Saunders admitted, "I was anxious to form my own impression."[3]

Negative feelings about Hynek had been all too common even before the swamp gas fiasco, and he freely admitted as much in an account of a return trip to Socorro a year after the Zamora sighting. In a 1965 report in the CUFOS files, he said, "There is . . . the opinion expressed on Socorro, and expressed to me a number of times in the past, by several people, that I am merely a part of a super-smoke screen . . . and that the whole Project Bluebook is a grand cover-up for something the government does not want discussed. Best way to give a lie to this, of course, is to point out that if this were the case, the U.S. government should also have been responsible for the sightings in France, Brazil, Spain, and in England. Maybe the U.S. government has really gone global!"[4]

That was before the swamp gas case. After the swamp gas case, well, this choice diatribe in response to an article in the August 29, 1966, *Washington Post* should be sufficient to illustrate Hynek's position in certain UFO circles:

> *Note here that Gobbledegook Hynek gives no hint of the government's 18-year "silence policy" on Flying Saucers, nor of the fat research grants by which the Air Force has hired him as its official civilian mouthpiece. Under instructions from his Psycho-Political Warfare boss in the Air Force, Hynek has placed the blame for all of the Flying Saucer "confusion" on the civilian scientists. Perhaps this astronomer artist in flumduddery had hopes of heading the $300,000 UFO research project . . . if so, his hopes were dashed on Oct. 7 when the Air Force announced . . . that the University of Colorado would undertake the 18-month, $300,000 study of reports of Flying Saucers . . . Perhaps this is because Hynek sacrificed his scientific credibility when he belched forth that "swamp gas" explanation of Michigan Flying Saucers last April.[5]*

At the other end of the spectrum, Hynek's October letter to *Science* magazine garnered enthusiastic support and demonstrated that at least some serious scientists and thinkers considered UFOs a subject of legitimate interest:

> *Dear Dr. Hynek: As a member of AAAS and, more importantly, a member of the scientific community, I wish to thank you for your recent letter to* Science. *Your letter stated concisely a number of things that needed to be said about "UFOs."*
>
> *I am not a "true believer" . . . This whole thing is an open issue to me. I have been dismayed, alarmed and confused by the hysterical, medieval, emotional orthodoxy of most of our brethren. The very idea of new phenomenon, normally an exciting one to anyone purportedly interested in knowledge, in this case is made a vehicle for jokes, vicious insinuation, etc.*

I have, to about 3 or 4 people, made slurring comments
about your own work in this area. I hereby apologize profusely
for accepting newspaper accounts rather than your own words.
I shall undertake retractions and corrections where required to
clear your good name.
   —*Faculty member, Lehigh University, October 31, 1966*

Which Hynek would be testifying before the committee in Boulder? The "artist in flumduddery"? The forthright scientist? The mystic reaching beyond the limits of science? Or someone altogether different?

Hynek's presentation began innocuously enough, with a lengthy overview of the air force's attempts to study the phenomenon. "The goateed professor of astronomy told us that the Air Force had originally gotten involved because flying saucers presumably fly—and anything that flies is the Air Force's problem," Saunders reported.[6]

Hynek went on to describe how Blue Book's minuscule budgets hampered its ability to do its job. The air force, for example, would seem to have an adequate supply of aircraft at its disposal, and yet Hynek and other Blue Book investigators were required to travel on commercial flights, and only then after waiting for approval. Because of this, they often arrived at the scene of a sighting several days after the event.

According to Hynek, Saunders wrote, "a brass curtain surrounded even Blue Book, with all the policies being set by the military rather than by the scientists."[7] Calling the project's reporting procedures "a numbers game,"[8] Hynek explained to the committee how UFOs listed as "possible aircraft" or "possible balloon" on the initial report would be mysteriously changed to "aircraft" or "balloon" in the year-end report. And because the multitude of project chiefs rotated in and out of the Blue Book command post with alarming frequency—and ran hot or cold on any given day, week, or month—the "unknown" cases were never consistently investigated and often simply fell through the cracks.

Hynek invoked the story of Johannes Kepler and Tycho Brahe

to explain why he continued to put up with the most repellant of his bosses—Captain Hardin, the aspiring stockbroker who was bored by UFOs; Captain Gregory, the "zealous UFO debunker";[9] and the perpetually enraged Major Quintanilla—and remain on the project: "Tycho Brahe had the data and Kepler wanted them," he explained to the committee, "and he knew that he had to get along with Tycho to get the data; and some of those cases interested me."[10]

After the formal presentations were over, Hynek and Vallee remained in Boulder for a second day, holding informal talks with members of the committee, who recognized the value of the two scientists' knowledge. Hynek was impressed with the enthusiasm and professionalism of the committee members and returned to Evanston feeling that good things would come from the trip. "As he continued to monitor what he termed 'the noise level' (the thousands of sighting reports), Hynek became increasingly more convinced that there was a signal hiding in all the noise," Saunders related. "He hoped an investigation such as ours could ferret out the best cases and increase the signal-to-noise ratio."[11]

Encouraged by his time before the Condon Committee, Hynek was beginning to recognize that a whole new attitude toward UFO research was in order. "In the fall of '66 was the real time I changed," he told Vallee. "I said, 'I'm going back and relook [sic] at this material with a different viewpoint, no longer assuming that the chances are strong that it's all a lot of junk. I must take the viewpoint that after all these years, the data may be genuine, as poor as it is (because it is not hardcore data, not the sort of data that a physicist wants). I'm just not going to continue to call all these people liars, deluded brains, and so forth, even though some of them can't tell Venus from a hole in the ground; they're people whose testimonies would be accepted in a court of law in any other context. By what right do I have to continue to doubt their words?'"[12]

Hynek's decision to revisit puzzling cases from the past couldn't have come at a better time, as there were now fewer and fewer new cases coming his way from Blue Book. One reason for this of course was that the Condon Committee was taking over much of the burden of investigating new cases, leaving Blue Book with relatively little to do. Another reason was that Major Quintanilla was still angry with Hynek over his actions at the recent congressional hearing. In one October 1968 letter, Quintanilla informed Hynek that "under no circumstances will you review the findings of the Condon Committee as an official Air Force Consultant," then he scolded Hynek for taking part in a panel discussion of UFOs at the National Electronics Conference the previous month: "I would appreciate it," he wrote, "if you would refrain from identifying yourself as an Air Force consultant when participating in pseudo-scientific panels of this type."[13]

Hynek's response was classic: "Your reference to 'pseudo-scientific panel' in my opinion is illogical; a panel could only be called 'pseudo-scientific' if its members were 'pseudo-scientists.'"[14]

Worse yet, Hynek also found himself blocked from the crucial database in Blue Book's files, a situation he described as "childish."[15] He was no longer given unimpeded access to the case reports, and when he was given access he was forbidden from making copies of files, even when he offered to supply his own copier paper.

Meanwhile, as he studied the hundreds of files that were shoehorned into his home office, looking for data that could help him make sense of the "unknowns," Hynek began to visualize a system by which he could categorize UFO events. Now enamored of the idea of finding the signal in the noise, the order amid the chaos, it occurred to Hynek that if he could place cases in discrete categories derived from the reliability of the witness and the nature of the sighting, then perhaps, over time, patterns might begin to emerge.

He sent up a trial balloon about his thinking in 1967, in a magazine that had, arguably, an even more avid readership than did *Science*. His article "The UFO Gap" appeared in the December issue of *Playboy* and became one of the most talked about articles the magazine had ever published.

Although the article was ostensibly a cautionary tale about the secret efforts of the Soviet Union to crack the UFO case before the Americans did, Hynek used the opportunity to describe his embryonic work in classifying UFO events. He seemed to hope that sharing his theories in *Playboy* would give him the best opportunity to influence the all-male Condon Committee. "At least, publishing in *Playboy* does give one a very considerable circulation," Hynek wrote to John Fuller when the article came out.[16]

"The existing evidence may indicate a possible connection with extraterrestrial life, the probable existence of which is generally accepted," Hynek wrote in *Playboy*. "If such life does exist and if there is any possibility of establishing communication with it, our scientific knowledge of that life might even be critical to our survival. Now let us be clear: The existence of extraterrestrial intelligence and the UFO phenomenon may be two entirely different things. But the latter, in itself, poses an interesting scientific problem. How can it be studied? Do we ignore it simply because the evidence we have does not follow the strict rules of scientific evidence?"[17]

How to remedy this situation? "UFOs, obviously, cannot be studied in the laboratory," Hynek pointed out to his readers. Neither, however, can a tornado, but this doesn't prevent atmospheric scientists from studying the *results* of a tornado in a lab. So accepting that a scientist could study UFOs based only on the results of a sighting, Hynek suggested that the Colorado investigators analyze cases based on two qualities: the strangeness (Σ) of the event, meaning "the difficulty in ascribing a simple scientific explanation for the report," and the credibility (C) of the witnesses, meaning the social and psychological standing of the person or persons reporting a UFO. If, over time, it was possible to identify concentrations of high ΣC cases, the Colorado group could focus its attention on the area in which the phenomena were occurring.[18]

If such an effort came up empty after a year, that, in itself, would be valuable information, but Hynek doubted that the effort would yield nothing. "To the contrary," he wrote, "I think that mankind may be in for the greatest adventure since dawning human intelligence turned outward to contemplate the universe."[19]

Whether or not the message made it through to the Condon

Committee, the assistant editor of *Playboy* informed Hynek that his article was a hit among the magazine's readers. "I should assure you that we seldom get this many letters about an article," he wrote. "Most of the mail, of course, wants to know whether to wear an ascot and how to get in touch with the latest Playmate."[20]

As popular as he was in the media, Hynek still had a job to do at Northwestern University, and Northwestern did not always like it when a chairman of one of its departments got too much of the wrong kind of publicity.

Not even the possibility of a big government grant could ease the discomfort felt by Northwestern's administrators toward Hynek's UFO activities. When Hynek suggested submitting a proposal for a $250,000 air force contract to transfer the Blue Book UFO reports into computer files using the university's resources, he quickly learned where he stood. "Because of the sensitive nature of any research related to UFO's we feel that for the contract to be negotiated by Northwestern University the following conditions must be met," wrote Dean of Sciences John Cooper in an August 1966 letter to Hynek. Cooper went on to list a series of difficult, time-consuming, nonessential conditions, not the least of which was the formation of a scientific board of advisors. That's right: Northwestern was requiring the formation of a scientific board of advisors to oversee a data entry project.

Hynek asked Cooper, "What wrong could there possibly be in reducing the Air Force data to machine readable format?" After threatening to take a leave of absence from Northwestern and take the contract to another university, he let loose: "I think the University is afraid of its own shadow . . . Is Northwestern afraid to do a scientific job that needs doing because of a little possible criticism? Are we a University or an ostrich farm?"

Hynek was outraged. He had been delivering prestige and money to the university ever since he came on board in 1960. Even with the cancellation of Project Star Gazer, Hynek's ability to fund and build not one but two new observatories for the university should have earned him a bit more gratitude from his superiors at Northwestern.

AMONG HIS OTHER ACCOMPLISHMENTS, Johannes Kepler discovered the last visible supernova in our galaxy, in 1604. It was only the third supernova visible to the naked eye in the Milky Way in the last thousand years or so. If you surmise from this that supernovae are rare occurrences, you are correct.

It is remarkable, then, that the first observatory that Hynek had built for Northwestern University was so well suited to finding these elusive exploding stars that he, his staff, and his graduate students discovered a record number of new supernovae in neighboring galaxies.

The Corralitos Observatory first gazed upon the universe in late 1965, when it was opened with great fanfare in the Rough and Ready Hills not far from Las Cruces, New Mexico. Initially, Hynek used the Image Orthicon system at Corralitos in an attempt to capture images of strange manifestations of light and color on the moon known as "transient lunar phenomena." The exercise was perfectly suited to the imaging system's accelerated light-gathering capabilities, but in the end it was unsuccessful. The Image Orthicon system excelled, however, when coupled with an automated search protocol and used to detect supernovae.

What makes a supernova super? Simply put, a nova is a short-term flare-up of a star caused by stellar material from a nearby star falling into it, while a *super*nova is a massive explosion caused when the core of a dying star collapses upon itself. It is the elemental cauldron of the universe, tearing the basic building blocks of creation apart in a violent cataclysm and hurling them off into space where they may in time seed new stars, planets, and even life. Supernovae are vital to our understanding of the universe, and the more we can find, the more we can learn about our past, our present, and our future. Hynek wanted to find more.

"In the 1960s, a first step to an automatic [supernova] search program was begun by researchers from Northwestern University. Astronomers J. Allen Hynek, William Powers, and Justus Dunlap constructed an automated 24-inch telescope at Corralitos Observatory . . ." wrote physicist Laurence A. Marschall. "The telescope moved under computer control, but observers sitting in

Dr. Hynek prepares for a long, chilly night at the telescope. *(Photo by Herb Comess. Courtesy of Northwestern University.)*

a room nearby watched the images of galaxies on a TV screen, comparing what they saw with photographs of the target galaxies. 'We were able to check a new galaxy every minute,' recalls Powers, 'so that on a long winter night we could check hundreds. Operators worked 2-hour shifts—it was grueling.'"[21]

In later years, Bill Powers remembered the project a bit more fondly: "I designed the computer programs and the telescope drive systems and all of that, and [Hynek] got the funding for it and the organizational backing," he said. "It was built, and it started discovering supernovas faster than anybody had ever done before. I think we saw in only the course of about two years fifty supernovas.

"We observed hundreds of galaxies with this telescope under computer control," Powers went on. "It would go from one to the other to the other. These poor graduate students had to sit and look at the comparison picture and the present picture and look for a change, because no mechanical way of doing it was good enough . . . So these poor guys spent thousands of hours, and they found fifty supernovas! We got on the telephone and alerted the observatories on our list, and they would get out their telescopes and take spectra; we gathered a wealth of information about what a supernova looks like just after it explodes."

The Lindheimer Astrophysical Research Center (LARC) was observatory number two, and it came about in a much different way than did Corralitos. Hynek's son Paul recalled that the genesis of LARC was much more politically fraught.

"I remember LARC being built at Northwestern," Paul said. "The Lindheimer family donated the money and my dad said, 'Okay, I know a good place in New Mexico, at Corralitos.' And they said, 'No, we want it in Chicago.' And he said, 'But the night lights in Chicago are such that it won't be very good.' And they said, 'It will be in Chicago.'"

LARC, built for $1.5 million on a landfill along the Lake Michigan shore, was surely an impressive sight, with its twin brilliant white one-hundred-foot domed towers and exoskeletal frame. And its accoutrements were of the highest quality, including a forty-inch telescope financed with the generous assistance of the A. Montgomery Ward Foundation and a sixteen-inch telescope funded in part by the National Science Foundation.

But, as Hynek had predicted, its proximity to the lights of Chicago made LARC all but useless for serious research. As a teaching facility, however, it was light-years ahead of the Dearborn Observatory, and on aesthetic grounds alone, it was a stunning sight on the Northwestern campus. It was noted as a structure "outstanding in its class" by the Chicago chapter of the American Institute of Architects and received an "Honor Award" from the Chicago Association of Commerce and Industry when it opened in 1966.*

In light of Hynek's proven abilities to bring such honors and riches to the university, it's difficult to understand Northwestern's reluctance to let Hynek take on the air force's data entry project. Perhaps it could be seen as a case of Hynek's "committee complex" writ large, for outside the context of academia, Chicago was effusively supportive of Hynek's UFO work.

In a December 22, 1966, feature article in the Cityscape section of the *Chicago Sun-Times* titled "The Flying Saucer Man," reporter Richard Lewis painted a sympathetic, affectionate, at times even heroic portrait of a man whose work "challenged the vitality and flexibility of the American scientific establishment"[22] and led directly to the formation of the Condon Committee. "More than any other man in America, Hynek is responsible for this first, tentative effort of science to meet the challenge," Lewis wrote.

Lewis went to some lengths to illustrate Hynek's contented family life, which had changed considerably with his wife Mimi's

---

* In part because of its site on the damp, windy shore of Lake Michigan, LARC aged quickly and was demolished after less than thirty years. Because of the extensive use of lead-based paint and asbestos in the structure, renovations to the building were judged to be cost prohibitive. LARC's exoskeleton proved so robust, however, that the explosive charges meant to bring down the towers only caused the structure to list slightly to one side. The two telescopes were donated to the Lowell Observatory in Flagstaff, Arizona, the site from which Percival Lowell first observed his "canals" on Mars.

entry into local politics and with the births of sons Paul and Ross after the move to Evanston. Lewis particularly enjoyed a story involving the two youngsters and a prototype toy. It seems a few days after Hynek's article had appeared in the *Saturday Evening Post,* he got a phone call from a toy manufacturer who made an interesting offer: "Would Hynek evaluate a toy flying saucer?"[23]

"No, Hynek said, that was out of his line. But he could refer the project to a couple of experts on toys, his youngest sons, Paul, going on five, and Ross, three and a half. The toymaker took him up on it and promised to send one," Lewis wrote. "No, two, Hynek said. An experienced father, he didn't want his household torn by civil strife over only one flying saucer."[24]

After noting that Hynek had spent his career avoiding the paper-pushing existence of a university dean, Lewis described Hynek's office at the Lindheimer Astrophysical Research Center as the place where Hynek "hides out" from "the hurly-burly of modern astronomy and associated research projects." The reporter observed that the office had no phone, but that it did have a remnant of Hynek's youth, his "Rosebud" perhaps: "an old yellow oak office desk, complete with a missing file drawer." The desk, an eighth-grade graduation present from his parents, Joseph and Bertha, had seen him through some forty-three years of life, moving with him from North Lawndale to Delaware, Ohio, to Baltimore to Boston to Evanston. "I can't remember how many times I've taken it apart and packed it for shipping," Hynek told Lewis.[25]

"I wish I could find the so-and-so who swiped the file drawer," he added. "It has sentimental value only."

At the conclusion of the profile, Lewis gave the last word to yet another man of science who felt he owed Hynek a debt of gratitude for having written his UFO letter to *Science* magazine: "Hynek's letter makes me feel better. As a fishery biologist, I have almost felt ashamed that I, too, have seen a flying saucer. I feel now that Hynek has given the scientific observer freedom to talk about these crazy flying machines."[26]

CHAPTER 16

INVISIBLE AT LAST

DESPITE THE FACT that so many institutions of higher learning had declined the air force's $300,000 UFO study, there were certain academics at those colleges and universities who were secretly interested in studying UFOs and very disappointed that their employers had turned their backs on that prize. Recall that when Hynek conducted his cloak-and-dagger interviews of professional astronomers around the country in 1952, he found that 41 percent of them were willing to take part in UFO research, on the condition that they be given intellectual cover.

The creation of the Colorado Project, then, gave rise to an emerging market for scientists interested in studying the UFO phenomenon. UFOs may have been leprosy to colleges and universities, but some of the employees of those institutions were seeking out alternative avenues by which they could safely explore the topic. Hynek, of course, had already been developing such an avenue, and he made it available to any and all interested scientists and thinkers.

The embryo of Hynek's "Invisible College" had taken shape when Bill Powers and Jacques Vallee had found their way to the Dearborn Observatory and begun working with him in the early 1960s. It grew a bit with the addition of Fred Beckman, an electroencephalography expert from the University of Chicago whom historian David Jacobs described as "a conspiracy-type guy." Of course, the group had neither a name nor a structure at first; it was, on one level, just a few men working together (with the exception of Beckman), operating a university observatory and sharing an interesting hobby, but it was, on another, a consortium of curious, rebellious minds determined to upset scientific orthodoxy. They just didn't quite know it yet.

"I was introduced to this informal group over ten years ago, when Dr. J. Allen Hynek . . . invited me to apply my background in computer science to a study of the statistical procedures used by Project Blue Book," Vallee wrote in 1975. "In the ensuing years, I learned much about UFOs which was not then, and still is not now, public knowledge."[1]

"Oh, it was very interesting," Powers said wistfully about the College. "We kept it separate from our scientific work. It was something we did on the side." Considering that Hynek would soon be sending Powers out to investigate sightings, the boundary signifying "on the side" must have been very blurry at times.

"I examined the 10,000 reports contained at that time in the files of the USAF, spending four years in sorting the signal from the noise . . ." Vallee wrote. "I found that frustration concerning this baffling problem was as high or even higher among military personnel as it was among the best informed of my scientific colleagues."[2]

Powers described the group's operating philosophy thusly: "There is something there. [We] don't know what it is, but there's something there. It's too widespread; there are too many cases where there are multiple witnesses to things we can't explain. Yes, people are ignorant; yes, they misinterpret ordinary things seen under unusual circumstances and all, but there it is. They're reporting something they saw."

The group was very open-minded about the true nature of UFOs and was willing to entertain almost any theory, up to and including something that Hynek had rejected back in the early days of Project Sign: the extraterrestrial hypothesis.

"To say that our star—the sun, is the only one to have planets will be akin to saying, for instance, that your cat can have kittens and no other cat in the world can have them," Hynek said, adding, "I am perfectly willing to admit the reality of the ETI's (Extraterrestrial Intelligences)."[3]

Decades later, there was a singular case that still stood out in Powers's memories as epitomizing the scientific interests of the Invisible College.

"We started getting reports of an object appearing in the northern sky. It grew in size until it was rather large, and then it moved around and then it faded away," he said. "But half of the people reported it was rushing at them; not just that it got bigger, but that it was rushing toward them."

The reports came in from Illinois, Wisconsin, Michigan, Indiana, Minnesota, and North Dakota, and the sightings were as startling as they were widespread. Some pilots were sure the light was flying alongside their planes, while one was certain he had seen it touch down on a runway and then take off again. One motorist panicked along with his three passengers and tried to escape the light by driving 50 miles an hour *backward*, thinking that *The War of the Worlds* was finally, really starting.

The object was "seriously scaring" witnesses, Hynek recalled, but "the Air Force had 'explained' these reports with a single interpretation: the star Capella."[4] Hynek doubted this. He happened to know Capella well, as the star is notorious for the way it appears to jump around and to flash different colors when it's close to the horizon.

Certain there was something else going on, Powers collected the phone numbers of the witnesses from the police stations, newspaper offices, and air force bases that had reported the sightings, and he started calling the witnesses back. "Powers obtained thirty new reports involving seventy-four people . . ." Hynek wrote. "They certainly had not all been observing Capella!"[5] That is, not unless Capella appears as a small point of light and then quickly expands until it's larger and brighter than the full moon, changing color from red to blue to green to silvery white as it does so, as many of the witnesses described it.

The light eventually "elongated and disappeared," according to the reports. A few minutes later, as witnesses across the entire Midwest struggled to catch their breath and one motorist finally felt safe enough to shift his car back into drive, a *second* huge ball of light appeared and the very same sequence of events happened *again*.

"I just started writing things down and plotting it all on a map," Powers recalled. "And, sure enough, all the lines of sight they reported converged somewhere over Hudson Bay. Oh, that's a long way away! Of course the baselines were pretty long, too; they were hundreds of miles apart, so we had a pretty good triangulation on this thing."

The explanation that eventually emerged proved how easily people can be deceived by their own senses and demonstrated that different branches of the armed forces really needed to keep one another better informed of their activities.

"There was a test!" Powers explained. "[The army] was sending up a sounding rocket with a payload of heated barium, which was released into space. They wanted to see how it would be excited by the electrons of the Van Allen belt out in space. That would give them some indication of how strong the radiation was up there. It was a valid scientific experiment."

In fact, the army had launched two rockets that night from Fort Churchill in Manitoba, Canada. They were launched minutes apart and rose to an altitude of 250 miles before deploying their payloads, explaining the second occurrence. Powers was delighted with the simple solution: "Now we had a case where people were reporting something that we knew had happened! This was 'an unusual thing seen under ordinary circumstances.' And, by golly, the observers came through that with flying colors. We were able to get data together that made sense; the correlations between the data from the different observers were what they should have been, and the observers were all ages and all occupations."

Despite the somewhat unexciting denouement, what the incident taught Hynek, Powers, and Vallee about witness reliability and the psychological dynamics of a UFO report was worth its weight in gold. "In that particular case the 'expert' was the guy who was wrong," Powers exclaimed. "He was the pilot of a commercial airliner. He thought it was something on a collision course with them and he took evasive action. He took evasive action, and it was two hundred miles away from him! So it turned the whole prejudice thing upside down. The guy who was driving home drunk from the bar gave a pretty good report of it: he got the direction, the time, the appearance. He was pretty much in agreement with everyone else."

This came to typify the differences between the Colorado Project and Hynek's Invisible College. Hynek and his colleagues understood the phenomenon, of course, but they also had a deep understanding of and sympathy with the human element. Condon's people didn't really seem to understand much of anything, and kept turning in reports like these:

- "This unusual sighting should therefore be assigned to the category of some almost certainly natural phenomenon which is so rare that it apparently has never been reported before or since."

- "In summary, this is the most puzzling and unusual case in the radar-visual file. The apparently rational, intelligent behavior of the UFO suggests a mechanical device of unknown origin as the most probable explanation of the sighting. However, in view of the inevitable fallibility of witnesses, more conventional explanations of this report cannot be entirely ruled out."

- "This is one of the few UFO reports in which all factors investigated, geometric, psychological, and physical appear to be consistent with the assertion that an extraordinary flying object, silvery, metallic, disk-shaped, tens of meters in diameter, and evidently artificial, flew within sight of two witnesses. It cannot be said that the evidence

positively rules out a fabrication, although there are some
physical factors such as the accuracy of certain photo-
metric measures of the original negatives which argue
against a fabrication." Final verdict: "fabrication."

Hynek and his colleagues quickly grew frustrated with the committee's methods and findings. Instead of looking back at Blue
Book's hundreds of perplexing unknown cases, as Hynek had expected the committee to do, Condon and Low decided that the
group would investigate only those cases that were reported during
its expected two-year lifespan. Moreover, the committee's investigators would not conduct any initial screenings to weed out obvious mistaken identifications and hoaxes, which would undoubtedly
make up the bulk of the reports. Those cases would be allotted time
and resources equal to those given to the relatively few true "unknowns." The truly puzzling cases—in other words, the high ΣC
cases most likely to reveal genuine scientific evidence—would, as
they were with Blue Book, be relegated to the sidelines.

It wasn't that the members of the committee weren't all earnestly trying to make the project a success. The problem was
that they seemed at times to be aiming in completely different
directions. There was constant communication among committee members and many long meetings in room 203 of Woodbury
Hall, lasting until the blackboards were crammed with lists and
every last stick of chalk had been worn to a nub. But there were
also many impromptu meetings made up of only two or three
colleagues, who sometimes neglected to share their own lists and
decisions with other members of the project.

One scientist wanted to study whether UFOs might be made of
plasma, another wanted to develop the ultimate UFO camera, yet
another wanted to create a psychological profile questionnaire for
UFO witnesses. The psychologists didn't understand the physical
scientists. The physical scientists didn't understand the engineers.
The engineers didn't understand the astronomers. All needed a slice
of the $300,000, and all felt their work deserved the biggest slice.

"The key word was 'scientific,'" admitted Saunders, "and each of us had his own idea as to the meaning of that all-important word . . .

"[T]hat which we lacked could not be given to us," he continued. "We had to discover, for ourselves, whether or not we could communicate across the boundaries of our individual scientific specialties and whether we could learn to accept and respect the judgments of our colleagues as the Project progressed."[6]

The project wasn't so much "ivory tower" as it was "Tower of Babel," and it did not take long for the tower to start crumbling. Despite its stated mission of assessing whether the UFO phenomenon was worthy of scientific investigation, the committee became mired in the much more entertaining—but essentially impossible to answer—question of whether UFOs were extraterrestrial spacecraft being flown around by aliens. Deep divisions among staffers erupted into open warfare: a poorly written memo from Robert Low made it sound as though the entire study was rigged, and psychologist David Saunders and electrical engineer Norm Levine were purged from the project in a very public and humiliating manner after criticizing Low's unfortunate memo. As for Condon, his scornful attitude toward UFOs, or, as he pronounced it, "Ooo-Fos," inflicted deep damage to the committee's credibility. And because the Colorado Project had been launched with such fanfare (and with such a lofty price tag), every stumble, every squabble, every defection was reported in the daily paper and on the nightly news in a most spectacular fashion. Amid the chaos, nearly everyone's professional reputation took a whacking.

It's no wonder that the subtitle of Saunders's 1968 book about his time on the Colorado Project was *Where the Condon Committee Went Wrong*. To Saunders, the breaking point came in late 1967 when Condon's comments made it increasingly clear that he had preordained that the final report would find no scientific value in continuing to study UFOs. Condon further admitted to Saunders that if evidence of the physical reality of UFOs *was* to be found, he would not release that information to the public. "The battle lines were now clearly and openly drawn—Condon and Low against the rest of us," Saunders wrote.[7]

"The Committee never worked as a coherent body and was torn by much internal strife," Hynek wrote with impressive restraint, although he went on to suggest that Dr. Condon had "throttled human curiosity at its source."[8] He and Vallee had to both consider themselves fortunate not to have been any more deeply involved in the project than they were.

The committee issued its final report in November 1968, with the approval of the National Academy of Sciences, and called time of death on Project Blue Book. Although the full report allowed some latitude as to the possibility of the physical existence of UFOs, Condon's summary—the only chapter that mattered to the air force—simply ignored the body of the report. Condon, on his own, recommended that, after collecting 13,134 UFO sighting reports, the air force should terminate Project Blue Book and get out of the UFO business for good.

Hynek was blunt in his review of the Condon report: "Final judgment of the work of the Condon Committee, which was not a study of truly Unidentified Flying Objects, but largely of easily identifiable objects, will be handed down by the UFO phenomenon itself," he wrote. "Past experience suggests that it cannot be readily waved away."[9]

It was hardly a surprise to Hynek, then, that after two years of drama and disappointment, Condon had reached a negative verdict, but it might have surprised Condon that Hynek came to see his committee's final report in a positive light.

Hynek's unexpected conclusion came from a chance encounter he had with French space scientist Dr. Claude Poher about a year after the committee shut down. The two met while Poher, who worked for the French government's NASA equivalent, was conducting rocket experiments at Cape Kennedy. "In the course of our conversation, he expressed a very serious interest in the UFO phenomenon," Hynek said, "and I asked him whence his interest sprang. He replied, 'I read the Condon Report.'"[10]

Hynek could not believe what he had just heard. How, he asked Poher, could anyone read the Condon report and come away wanting to know more about UFOs?

"[Poher] replied in a most serious manner, 'If you really *read* the Condon Report and don't stop with Condon's summary, you will find that there is a real problem there.'"

Hynek agreed, writing later that the Condon report, taken as a whole, could only lead the open-minded reader to one conclusion: "there should have been no question as to whether UFOs are a fitting subject for further scientific inquiry."[11]

Project Blue Book wound down quietly a year after the release of the final report, and a great calm settled over the issue. "To the majority of the public this was indeed the *coup de grâce* to the UFO era. Science had spoken," Hynek wrote. "UFOs didn't exist, and the thousands of people who had reported strange sightings (and the probable many thousands more who were reluctant to report) could all be discounted as deluded, hoaxers, or mentally unbalanced."

With the demise of Project Blue Book, Hynek found himself suddenly liberated from the frustrations, compromises, and bullying of the air force. He was a free man.

This worried his friends.

Bill Powers, who enjoyed good-natured debates with Hynek about the nature and meaning of UFOs, was about as open-minded as anyone. Yet when Hynek's newfound intellectual freedom led him down some unexpected paths, Powers was less than accepting.

"[Allen] got involved with this guy named Ted Serios, who was doing 'spirit photography' with a Polaroid camera. He was taking pictures of things from different parts of the earth and different eras of time," Powers recalled. "Well, I think [Allen] got hooked on that one."

"Allen dabbles in too many things," a worried Vallee recorded in his journal. "A few months ago he was fascinated with psychic surgery. Now he is publicly quoted as supporting an even more shaky affair, the alleged 'psychic photography' of Ted Serios, a beer-drinking 'psychic' who stares into a Polaroid camera and produces pictures of the leaning tower of Pisa and other monuments."[12]

When Powers called on a magician friend to demonstrate to Hynek how Serios might be faking his spirit photographs, Hynek was not amused. "He blew up in my face!" Powers said. "He said, 'Just because you can do it, that doesn't mean that's how Ted Serios does it!' So that's what I mean when I say I think he got hooked!"

Hynek's son Paul, meanwhile, recalled accompanying his father to psychic conferences and meeting Robert Monroe, the man who first shed light on the concept of the out-of-body experience. "[Monroe] wrote a book called *Journeys Out of the Body*," Paul said. "A fantastic, wonderful man. My father took me to three successive annual ESP conferences where he was a speaker, and I met all these psychics, and Robert Monroe—who's not really a psychic, but was in that kind of fringe science—you could see him and just know he was this hardscrabble guy who is not lying. He might be deluded in some way but he's not lying."

Powers and Vallee shouldn't have worried themselves over Hynek's "dabbling" in paranormal phenomenon, for they knew that such curiosity was an occupational hazard for a scientist with an alert and questioning mind. Why not at least consider the possibility of contact with the supersensible realm, whether it involved a Polaroid photo of the Leaning Tower of Pisa or a psychic voyage outside one's physical body? Why not at least *play* with the idea?

Didn't the same words Hynek had written about Kepler in college apply to him now? "He never lost his clear reasoning powers in metaphysical speculation," young J. Allen had once written about his hero; certainly, the same held true for him. After so many years of being constrained by convention, Hynek was allowing his imagination to toy with ideas and concepts that had been closed off to him since his days—and nights—as a student. The young mystic wondering out loud to Otto Struve what the world would be like if the human race evolved emotionally rather than physically was back.

FREEDOM FROM BLUE BOOK also meant more time at home. The three older Hynek children, Scott, Roxane, and Joel, were brought up in a household where their dad's work life was contained in a separate bubble from home life, as was common in so many households in the 1940s and 1950s. Thus, the highlight of everyone's year, the annual August visit to the family cabin, was a time to bond and celebrate and, at least in principle, forget about work. Even if Dad did spend a lot of his time in seclusion in his little office behind the main cabin, writing his books and listening to Russian radio, he would now and then reveal his true love for his family, albeit in surprising ways. Paul recalled a time when older brother Scott, who then had his pilot's license, flew to the cabin in a seaplane: "He was flying around and my dad got so excited running down the dock that he sprained his ankle real bad."

Back at the family home in Evanston, the older siblings seemed to have a way of leading their little brothers into mischief, which is a surefire recipe for happy childhood memories. "Reporters came over because we dug so many holes so deep in the backyard and found so many things that they'd come and interview us about that, [things like] farm implements," Paul recalled with a wistful glee. "My brother Joel hooked up a pulley system, so that we would dig so deep that we'd be filling up these buckets, then we'd start going horizontal. My mom had one simple rule, that if she came walking in the backyard and looked down the hole and she couldn't see us, we had to stop, fill it in, and start digging another hole." Dad must have been chuckling the whole time; for once, the reporters didn't want to talk to him!

Scott Hynek recalled in an interview how his dad's growing fame as a UFO authority created a unique family game to see who could get to the phone first when the inevitable call came during dinnertime: "Sometimes it was a guy calling himself Prince Michael of the Perseids. Another would warn that star movements were indicating that their father's life was in danger."[13]

"The phone would frequently ring at dinner," Scott said. "[Dad] would always take the call. A certain amount of these people were crazy, some weren't. It was hard to tell the difference. But if you're in your car and it suddenly stops and you see something in the sky, you're going to seem strange even to people who know you. He gave people a chance to tell their stories."[14]

In that same article, Paul revealed some details about his home life, claiming that his bedroom housed the world's largest library of UFO books, "much to my chagrin," and that every year the family Christmas tree was decorated with lights shaped like flying saucers. "We were a normal suburban family," he recalled. "But there were all these incongruous things, though."[15]

Hynek's home office, meanwhile, became a repository for UFO reports, Blue Book documents, and comic strips—Hynek's favorite section of the newspaper. Paul remembered that his dad's inner sanctum was filled with mementos of his world travels—"it was a mix of astronomy and world culture"—but the overall picture that emerged from recollections of family and friends was that the office on Ridge Avenue was too small a space overrun with too much unsorted, unfiled material. The congestion was so bad that Hynek's colleagues would wait until Allen and Mimi were away, then sneak into the house to sort and catalog the papers in his office. It was a job without end.

"THE YEARS IMMEDIATELY FOLLOWING THE [CONDON] REPORT WERE, paradoxically, a Golden Age of UFO research," wrote Swords and Powell. "The Colorado Project had awakened many academics and individuals, and they came, at least briefly, out of the closet with their interest."[16]

"I have positive evidence from personal correspondence and conversations with scientists that their interest is increasing but that it is still, in most cases, anonymous," Hynek wrote about this "golden age." "There is truly a growing 'Invisible College' of scientifically and technically trained persons who are intrigued by the UFO phenomenon and who, if provided with opportunity, time, and facilities, are most willing to undertake its serious study."[17]

One remarkable letter sent to Hynek in October 1968 bears this out. "Your colloquium has stimulated a great deal of thought about various aspects of identifying UFO's," wrote James M. Peek, a scientist at Sandia Corporation, "and I am sure that a number of people here, representing a great diversity of talents, are willing/anxious to matriculate in the Invisible College."[18] Note that Sandia is a high-security nuclear weapons development laboratory that was, at the time, operated by AT&T for the U.S. government, and that Hynek had blatantly broken Blue Book protocol to give what would seem to have been a strongly "pro-UFO" colloquium to Sandia staff. That there were "a number" of government scientists up to their elbows in our nuclear weapons research who were "willing/anxious" to participate in the Invisible College (Peek cc'd three other Sandia staffers on his letter to Hynek) is rather astounding.

Even before the final act of the Colorado Project, Dr. Carl Sagan of Cornell University decided that the Condon Committee provided a stellar opportunity for public education and showmanship. In 1968, Sagan and his pal Thornton Page persuaded the American Association for the Advancement of Science to hold a general symposium on UFOs at its annual meeting. The symposium was scheduled for the December 1968 meeting of the AAAS, but when it was learned that the Condon report would not be published in time for members to have read it, the group postponed its UFO symposium for a full year.

That fact alone transmits significant information. Had the symposium been held in 1968, *before* the Condon final report could be studied, the debate about the nature of UFOs would have been open and expansive, and the association would have been *advancing science* in ways Sagan would not have cared for. By waiting until long after the report was issued—*after* the Condon Committee had snuffed the UFO flame, killed Project Blue Book, and left UFOs in general disrepute—AAAS all but guaranteed that the symposium would be taking on a dismissive tenor and that the presenters calling for more study of UFOs would be starting from a defensive crouch.

A few months before the symposium, Sagan wrote to Hynek and bemoaned the "drift away from science" he was seeing in America's youth, and he expressed the hope that the UFO event might help reverse that trend. "The proposed symposium has no speakers without serious scientific credentials," he wrote. "We expect the symposium to be chaired both with a firm hand and with a sense of humor . . . I do not see how such a symposium can fail but to serve science well."[19]

Hynek wrote to Sagan later that he was looking forward to "a very interesting time" at the symposium. He explained that he was having a philosophy professor look over his paper and added that Thornton Page had written to him, expressing his hope that whatever Hynek said in his presentation wouldn't make Donald Menzel "sore."

"I had replied that if Menzel got 'sore' that was his problem," Hynek wrote, adding that "UFO reports continue to come in regularly in my personal mail and many of them are high ΣC cases—and the problem is just as alive as ever."[20]

Despite the delay, Hynek appreciated that the symposium was held at all. "As I look back over my past twenty-one years' association with the UFO problem," he said, "I note that the intellectual climate today is enormously better for taking a good look at it than it was even a few years ago. This symposium is itself an example: it would have been impossible to have held it a year or two ago."[21]

So it was happening. There was that.

"THERE ARE MANY IDEAS WHICH ARE CHARMING IF TRUE, which would be fun to believe in, which are a delight to think about," Sagan began in his presentation to the symposium.[22] He then listed those topics in no particular order: reincarnation, the search for the philosopher's stone, the pursuit of immortality, psychokinesis, precognition, telepathy, time travel, out-of-body experiences, and "becoming one with the universe."

Because these concepts were so "charming" and had such deep emotional appeal, Sagan said, "they are the ideas we must examine most critically."[23]

In what kind of science does the fun factor of a phenomenon determine the level of criticality with which it must be examined? Having so recently written to Hynek that the symposium "couldn't help but serve science well," one might wonder why Sagan would undermine his own prediction by taking such an intellectually dishonest position. Nonetheless, Sagan continued to dig, making his intentions and his opinions clear: "The idea of benign or hostile superbeings from other planets visiting the earth," he declared, "clearly belongs in such a list of emotion-rich ideas."[24]

Having so deftly shifted the topic of the symposium from UFOs to "aliens from other worlds visiting Earth," as he delighted in doing, Sagan moved on to the central point of his presentation, demonstrating how massively improbable it was that those aliens would be visiting Earth at all. He explained how calculating the odds of intelligent life developing elsewhere involved determining the tiny fraction of all the stars that have planets, the tiny fraction of those planets that are suitably located with respect to their star, the tiny fraction of those planets on which life emerges, the tiny fraction of those on which intelligent forms of life have arisen, and the tiny fraction of those which have evolved a technological civilization far in advance of our own.

He was not incorrect in saying that those calculations must be made. But it is, nonetheless, jarring to see the man who made his name by opening the eyes of the public to the majestic immensity of the universe in his TV series *Cosmos,* the man perhaps best known to the world for proclaiming to us time and again that the universe was populated with "billions and billions" of stars, arguing vociferously and mockingly that when it came to intelligent life visiting Earth, this cosmos—his cosmos—was defined by scarcity.

Is it "billions and billions" of stars, or is it "slim to none"? We know now that it's bigger than both. According to the astronomy department at Cornell University, where Sagan taught for nearly thirty years, "billions and billions" doesn't even come close to describing the number of stars in the universe. Cornell's online "Ask an Astronomer" service says, "For the observable Universe, it is estimated that there are as many as 200 billion galaxies, but we aren't able to see all of them as our telescopes are not sensitive enough."[25]

One of the founders of Cornell's information service more recently used a rough estimate of 10 trillion galaxies in the universe to come up with a truly staggering count of the number of stars. According to David Kornreich, now an assistant professor at Ithaca College in New York, the total number may be close to a septillion stars.[26] For those whose counting skills waver past the first few billions, that's a number 1 followed by 24 zeros, or about a third of a line of text in this book. And Kornreich admitted even that number is "likely a gross underestimation."[27]

If that strikes the reader as too speculative an approach, then consider this: in May 2016, NASA reported that one of its newest orbital space telescopes, this one named after the seemingly ever-present Johannes Kepler, had discovered more than 2,300 exo-planets (planets orbiting other stars). These were in addition to another 900 or so verified known exoplanets already on the books. The Kepler telescope, one of the successors to the Hubble Space Telescope, had, at the time of NASA's announcement, scanned more than 150,000 stars since it became operational in 2009. Of the more than 3,200 confirmed exoplanets, 21 were believed at the time to be "Earth-like" planets that could potentially support life, according to NASA.[28]

A simple calculation shows that approximately 1.6 percent of the exoplanets known in 2016 were believed to be "Earth-like." But the only thing "Earth-like" means is that a planet may spawn life similar to ours. Any number of non-Earth-like planets could spawn other types of life that could become intelligent and travel into space.

NASA's tallies have only climbed since, and they will continue to climb as time goes on and as our technology enables us to peer farther and farther into space. But the numbers will not simply climb; if recent history is any guide, they will multiply, then mushroom, then explode, to the degree that Paul Hertz, NASA's Astrophysics Division director, made the stunning statement that "thanks to Kepler and the research community, we now know there could be more planets than stars."[29]

In other words, today's technology, directly inspired by Project Star Gazer, tells us that there might be more than a septillion planets in the universe and that perhaps 1.6 percent of them are Earth-like and capable of supporting life similar to our own. And that may *still* be a low estimate.

It can be argued, of course, that Carl Sagan didn't know any of this in 1968 and thus can be forgiven his conservative calculations, but one might well wonder why such a visionary scientist didn't foresee it.

Sagan concluded with a comment that NASA's planetary probes and radio listening projects were the best way to contact extraterrestrial life, holding far more promise than did the study of UFOs. It was a sensible thing to bring up because Sagan was, at the time, the director of the Laboratory for Planetary Sciences at Cornell, and laboratories for planetary sciences at universities need funding from NASA almost as much as they need planets.

"There is much in the UFO problem to be astonished about—and much to be confused about, too," Hynek said when it was his turn to address the AAAS.[30] His speech, lengthy as it was, bears close scrutiny, as it showed the workings of a mind in transition. Hynek had still not found his footing as his two-decade-long career with the air force came sputtering to an end, but the text of his remarks reveals that he was using this moment of uncertainty to rethink old ideas at the same time he was thinking of new ones.

The basis of his speech was a conversation he had had with the Canadian philosopher of science Thomas Goudge on the matter of scientific methodology as it pertained to the UFO problem. Goudge, he said, was fascinated by how UFOs affect the manner in which science advances. "Goudge notes that throughout history any successful explanation scheme, including twentieth-century physics, acts somewhat like an 'establishment' and tends to resist genuinely new empirical observations, particularly when they have not been generated within the accepted framework of that scheme," Hynek said.[31] Goudge felt that this accounted for "the present establishment view that UFO phenomena are either not really scientific data at all (or at any rate, not data for physics) or else are nothing but misperceptions of familiar objects, events, etc. To take this approach is surely to reject a necessary condition of scientific advance."[32]

Hynek then quoted Dr. Erwin Schrödinger, the father of

quantum mechanics: "The first requirement of a scientist is that he be curious; he must be capable of being astonished, and eager to find out."

"Perhaps he should have added," Hynek said, "'and be ready to examine data even when presented in a bewildering and confusing form.'"

Bewildering and confusing described the fact that UFO reports were so consistent over time and throughout the world. For decades, witnesses around the globe had described seeing a very specific and limited type of object, predominantly flying discs, cigars, and orbs. "If there were some worldwide compulsion to report strange things, why are only these particular types of strange reports preferred from the infinite universe of all possible strange reports?"

Hynek then went on to describe a new type of UFO categorization system he had been developing, and in so doing he etched his name indelibly in history. First came the UFO events that took place at a distance—the daylight disc, the meandering nocturnal lights, and the visual/radar sighting. To those he added three categories of cases he felt offered the most useful and interesting data of them all: the "Close Encounters."

"I divide the close encounter cases into three subdivisions: the close encounter, with little detail; the close encounter with physical effects; and the close encounter in which 'humanoids' or occupants are reported," he told the group. Although the "physical effects" variant was the most appealing to him, he acknowledged that the "humanoid" encounters possessed their own uniquely repellant appeal. "This latter subgroup, of course, has the highest strangeness index and frightens away all but the most hardy investigators. I would be neither a good reporter nor a good scientist were I deliberately to reject data. There are now on record some 1,500 reports of close encounters, about half of which involve reported craft occupants. Reports of occupants have been with us for years but there are only a few in the Air Force files; generally Project Bluebook personnel summarily, and without investigation, consigned such reports to the 'psychological' or crackpot category."

"Although I know of no hypothesis that adequately covers the mountainous evidence," Hynek said in closing, "this should not and must not deter us from following the advice of Schroedinger [*sic*]: to be curious, capable of being astonished, and eager to find out."

CHAPTER 17

THE UFO EXPERIENCE

TWO AND A HALF HOURS.

That was how long it took for the entire human race to adopt a new belief system: 2 hours, 31 minutes, and 40 seconds is how long Neil Armstrong and Buzz Aldrin, Earth's first emissaries to another world, spent on the surface of the moon. On July 20, 1969, after a three-day voyage across space, their lunar excursion module (or LEM, perhaps a more refined version of the craft spotted by Lonnie Zamora in 1964) touched down in a flat, rocky basin called the Sea of Tranquility. From the moment the LEM's landing struts came to rest on *terra luna* and the blasting rocket was cut off, no doubt remained that humankind possessed the ability to travel safely to other worlds. Armstrong's "one small step" onto the dusty lunar surface became immortal for good reason: at the moment of humankind's grandest achievement in science, Armstrong reminded us that a single small act of a single small individual can change history.

Hynek, meanwhile, was taking his own "one small step," publicly subjecting the Condon Committee and U.S. Air Force to his subtle, deprecating wit. "The verdict has been handed down: Unidentified flying objects (UFOs) are figments of the imagination," he wrote in the *Christian Science Monitor*. "And the sentence has been pronounced: UFOs are henceforth banned from the land—nay, from the world—and no serious attention should be given them in the future."[1]

Hynek clearly felt that the future of UFO research depended on a strong public rebuttal to the Condon Committee's final report, and he did not hesitate to lead the charge. In his *Christian Science Monitor* article, he spelled out the reasoning Dr. Condon and his cohorts employed time and time again when confronted with a UFO case that couldn't be explained away as a natural phenomenon:

> *Investigators are faced, then, with at least three alternatives. (1) The witnesses suffered a major delusion; (2) an actual craft was present but answering to a higher level of physical laws than are known to our physical scientists; (3) no material object was present, but there was something there that gave all the impressions of being physically real and that could affect people, animals, and inanimate objects.*

The second alternative was not an option for the Condon Committee, as it called into question the very nature of the members' narrowly defined reality. This annoyed Hynek, who pointed out that "less than a hundred years ago they also would have ruled out categorically the possibility of nuclear energy, television and space flight." The third alternative could also be disposed of immediately, he claimed, "since it suggests the paranormal and thus akin to poltergeists, ESP, and the whole world of the occult."[2]

That left Condon and his committee, Hynek wrote, with the very palatable and popular "delusion hypothesis,"[3] which they wielded with abandon.

Hynek closed his article with the admission that science was in 1970 no closer to understanding the UFO phenomenon than it had been twenty years earlier. The admission was tinged not with defeat, however, but with a strange optimism:

"Persons with true scientific curiosity will watch with interest the coming post-Condon and post–Blue Book years," he wrote.

"Will 'incredible tales told by credible people' cease, now that the verdict has been handed down, or will some pilots and other people in highly responsible positions continue to have UFO experience?

"Study of the UFO phenomenon," he concluded, "will go on despite the ridicule barrier."

But in what form would it continue?

Despite the dawning of this golden age, some of Hynek's closest colleagues at the Lindheimer Astrophysical Research Center were going their own ways. Vallee had earned his Ph.D. in computer science and had taken a job at Stanford University. As NASA funding tapered off after the Apollo program, meanwhile, Powers had less and less to do at the observatory.

"The money went away and I decided I had to get back to real life," he said. "I was living a dream. As a kid I always loved telescopes and space and science fiction and astronomy, and here I was actually working in an observatory with a forty-inch telescope! . . . It was a real dream, but all good things must come to an end.

"I was actually designing an all-sky photometer for use on the moon," he said. "I got it finished and it was going to fly on Apollo 18. But you know the last Apollo was Apollo 17, so I missed the boat, or the boat missed me."

Looking for a more reliable boat, Powers took a position at the *Chicago Sun-Times,* where he stayed for nearly twenty years before pursuing his first love: the study of behavioral science.

While Powers moved on from UFO work, Vallee's close friendship and working relationship with Hynek endured, and he remained a vital force in the Invisible College for years. This despite Vallee's frequent frustration with Hynek, whom he now saw as "a quiet man, who disliked confrontation and scandal, feared authority and was in awe of secrecy."[4]

Hynek also maintained close ties with his former student and Blue Book assistant Jennie Zeidman. Since leaving Ohio State's employ in 1956, Zeidman had taken a job at Battelle—another unexpected consequence of Project Stork—and married a co-worker there. Based on their correspondence over the years, it can be said that Hynek shared more with Zeidman than he did with any other friend besides his wife, Mimi, Vallee included.

"Hi Jennie—60 today! Think I'll make it to Halley's comet?" Hynek wrote to her on May 1, 1970. "I expect to see it with you. Weren't we in some cathedral in England when I mentioned to you that Halley's comet had just 'turned the corner' and was on its way back (to get me!)?" Then, in the middle of thanking her for a book she had sent him for his birthday, Hynek casually mentioned that the air force had recently re-signed him as a consultant on a mystery project. "It may be Rooshian devices or it may be UFOs," he teased.

Zeidman was stunned: Only five months after the calamitous end of Project Blue Book, and Hynek was once again under contract to the Foreign Technology Division (FTD) at Wright-Patterson Air Force Base? How was that possible?

One clue could be found in a notice in the Northwestern University newspaper from 1977, which announced, "Professor J. Allen Hynek, a world-renowned astronomer and head of the UFO Studies Center will give a Fireside on Spy Satellites at 7 pm at the Foster-Walker Complex."[5]

When, exactly, did Hynek become an expert on spy satellites? When Hynek told Zeidman about the secret contract he'd signed with the FTD, he had hinted that the work might involve Russian "devices," and at that time U.S. and Soviet spy satellites were filling the sky, snapping pictures of the enemy's movements and activities and eavesdropping on the enemy's conversations from high orbits. It would certainly make sense for the air force unit tasked with identifying and understanding whatever our enemies might place in the sky to consult with the man who created the sciences of high-altitude observing and satellite tracking.

Whatever doubts she may have had about Hynek's new air force contract, Zeidman must have kept them hidden. In her reply to him a month later, she instead addressed Hynek's fear of Halley's Comet, which was, after all, still sixteen years away from Earth: "I wish you would get over your hysterical idea that The Comet is out to get you," she chided him. "You'll get yourself so worked up that you're apt to step outside to see it and fall down the back steps just to prove your point."

Zeidman had been one of the first to know that Hynek was writing a book about UFOs, and he looked to his old friend to offer editorial advice and encouragement as his manuscript neared completion. "Dear Allen," Zeidman wrote, "Your readers want you to level with them, or, putting it another way—write as if the subtitle will be <u>WHAT YOU'VE ALWAYS WANTED TO KNOW ABOUT UFOs BUT COULDN'T GET ANYONE TO TELL YOU.</u>

"Anyone who has ever bought a UFO book will buy one that says HYNEK on the front. (HYNEK and UFO <u>must</u> be the two biggest words on the cover)," she went on. "Your colleagues will buy the book IF the reviews say it is a <u>serious</u> monograph, and Quintanilla and Condon will <u>borrow</u> copies to see if there is anything they can sue you for!"

READERS OF HYNEK'S *THE UFO EXPERIENCE: A SCIENTIFIC INQUIRY,* published in 1972, were immediately put on notice that the UFO book they were about to read was far different from what they were used to.

"During my many years as scientific consultant to the United States Air Force on the matter of Unidentified Flying Objects I was often asked (and frequently still am) to recommend 'a good book about UFOs,'" Hynek wrote in his preface. "Very often, too, the request was accompanied by remarks along the line of 'Is there really anything to this business at all?' 'Just what's it all about anyway—is there any reliable evidence about UFOs?' or 'Where can I read something about the subject that wasn't written by a nut?'"[6]

"With a few notable exceptions I have been hard pressed to give a good answer to such questions," he added drily.

For the next few hundred pages, Hynek put his money where his mouth was, offering a full-throated rebuttal to the Condon report and a postmortem on Project Blue Book. In his own clever way, he wrote what he and many observers felt the Condon report *should have been.* First, he defined the phenomenon by delineating the various types of UFO events under examination, focusing on Close Encounters of the First, Second, and Third Kinds.

A Close Encounter of the First Kind is one in which a witness sees a UFO within five hundred feet or so, close enough to make out detail but not so close as to make physical contact. A Close Encounter of the Second Kind is one in which the UFO has a physical effect on the environment, such as scorching nearby plants, leaving strange markings on the ground, or causing a car's engine to stall. A Close Encounter of the Third Kind is one in which the witness sees and sometimes interacts with beings that appear with the UFO. He may not have realized it, but in an odd sense, he had created an analogous version of his hero Kepler's Three Laws of Planetary Motion.

In an interesting historical note, after Zeidman had an early look at the manuscript, she told Hynek that using the three Close Encounters categories as chapter titles was a bad idea: too many syllables. Couldn't he rewrite them, she wondered, make them catchier? Hynek considered her advice, but he liked the titles and would not change them. Thus is history made.

In later chapters, Hynek recounted in detail several truly puzzling "prototype" cases from each of his categories (noting that there can be some overlap between categories: a daylight disc first seen at a distance, for example, can also become a Close Encounter). He thoroughly analyzed the witnesses, their testimonies, the facts of the case, and how each element was investigated. Of course, he pointed out the flaws and shortcomings of each Blue Book and Condon Committee investigation.

And, along the way, he digressed into tales of triumph, tragedy, and comic irony that kept reminding the reader that these were portrayals not of witnesses but of "real everyday human beings with jobs and families."[7] Indeed, he had mentioned in a 1970 interview that he initially considered naming the book *UFOs as a Human Experience.*

"These people are not merely names in a telephone book," he wrote; "they are flesh and blood persons who, as far as they are concerned, have had experiences as real to them as seeing a car coming down the street is to others."[8] The people in his book were, in other words, flawed, frightened, overwhelmed, doubtful, indecisive, prone to errors of judgment, and terrified of embarrassment and ridicule.

Take, for example, the story of a UFO event that took place in 1968 in Victoria, British Columbia, outside a conference attended by hundreds of professional astronomers: "Word spread that just outside the hall strangely maneuvering lights—UFOs—had been spotted," Hynek wrote. "The news was met by casual banter and the giggling sound that often accompanies an embarrassing situation. Not one astronomer ventured outside in the summer night to see for himself."[9]

Later in the book Hynek tells his own tale of misjudgment, in which he admits his mistake in dismissing the 1957 Levelland, Texas, Close Encounters of the Second Kind as "ball lightning." He was, he confessed in the book, too busy dealing with the Sputnik crisis to be able to offer any kind of scientific analysis of the impressive event, in which fifteen people claimed to have seen a glowing egg that caused their vehicles' headlights and motors to die, so he vouched for Blue Book's simplistic explanation. Hynek's regrets concerning the Levelland event were especially acute, as Close Encounters of the Second Kind—sightings in which the UFO has physical effects on the environment—were to Hynek the most promising cases of all.

Then there was the heartbreaking case of Dale Spaur, prime witness in a Close Encounter event whose reputation and life, Hynek related, were all but ruined by the callous indifference of Hector Quintanilla.

Spaur, a deputy sheriff in Portage County, Ohio, was out on patrol with his partner Wilbur Neff on the night of April 16, 1966, when they came across what appeared to be an abandoned car alongside a highway and pulled up behind it to investigate. Spaur and Neff left their squad car to ascertain whether the vehicle was occupied, but first Spaur turned around to check that the road behind them was clear. That's when his troubles began.

The two deputies had shared a healthy laugh minutes earlier, when they heard a report on the police radio that a woman in a neighboring county had seen a brilliantly lit object the size of a house flying over her town. But when Spaur saw "this thing"[10] rising up from behind the trees and approaching him and Neff, he forgot to laugh. He stood frozen as the object drew closer, getting brighter and brighter every second, and told Neff to turn around to take a look. Neff had to look down, the object was so bright, and Spaur worried that his clothes might catch on fire.

"I was pretty scared for a couple of minutes," he reported; "as a matter of fact, I was petrified."[11]

The terrified deputies took refuge in their squad car and radioed in their situation. The radio operator at the station told Spaur to "Shoot it!" but Spaur was reluctant to do so: "This thing was, uh, no toy; this—hell, it was big as a house! And it was very bright; it'd make your eyes water."

Then it shot off. Spaur and Neff were ordered to follow "the thing," now moving quickly to the east. The pair sped off in pursuit, and forty white-knuckle miles later, Spaur and Neff were contacted by Officer Wayne Huston, who confirmed that their target was headed his way. He reported observing an object 800 to 900 feet in altitude, shaped like an ice-cream cone with a "partly melted down top,"[12] traveling in excess of 80 miles an hour. Huston pulled out behind Spaur and Neff and they followed the object, now streaking toward the Pennsylvania border at speeds over 100 miles an hour.

Thirty miles down the highway Officer Frank Panzanella of Conway, Pennsylvania, soon saw the object—a "half of a football"[13]—headed his way, followed shortly by two police cars approaching from Ohio at breakneck speed. The four officers met up and stood outside their cars and watched the object as it moved across the sky, then shot straight up "real fast," to about 3,500 feet. At this point a plane from the nearby airport flew under the object, after which the object shot straight up again until it "got as small as a ballpoint pen."[14] Panzanella was sharp enough to note that Venus was visible in the sky quite apart from the brilliant UFO—on the opposite side of the moon, in fact.

This was an ideal case: four credible witnesses "sequentially and independently involved,"[15] who saw the same thing at the same time, and whose descriptions matched in virtually every respect (and, if you count the woman who first reported the object to the police, the count rises to five independent, sequential witnesses). It was a natural for Hynek to investigate, but because the event took place so soon after the swamp gas incident, Quintanilla looked into it himself and shut Hynek out. Later Quintanilla would write that this was "the incident which convinced me that Dr. Hynek was a liability."[16]

Quintanilla decided for reasons of his own to interview only Deputy Spaur, only over the phone, and for less than five minutes. Spaur said that Quintanilla's first words were, "Tell me about this mirage you saw,"[17] and the interview went downhill from there.

"Quintanilla's method was simple: disregard any evidence that was counter to his hypothesis," Hynek wrote. Unfortunately for Quintanilla, Congress was now keeping a closer eye on Blue Book's activities in the wake of the swamp gas episode, and a local congressman made enough noise at the Pentagon that Quintanilla was forced to reexamine the case.

It is not difficult to guess how this made Quintanilla feel: "I wasn't at all happy."[18]

"This time the interview was long and involved," Hynek reported,[19] noting that Quintanilla spoke with both Spaur and Neff this time, as well as with the officer who had been the radio operator that night and the county sheriff, who served as a character witness. "However, it excluded two prime witnesses," Hynek wrote: the two officers who joined Spaur and Neff later in the chase were not asked to testify, a key omission.

Sadly, Spaur now knew what to expect from Quintanilla, so his tone throughout the second interview was guarded, hesitant, and angry. At one excruciating moment, when Spaur said he expected Quintanilla to describe his sighting as a hallucination, Quintanilla peevishly responded, "I'm an officer in the United States Air Force, and I don't call anybody a nut."[20]

Despite this newfound tolerance, Quintanilla's revised verdict was much the same as the original: Spaur had been chasing Venus. He was a nut after all.

When, some months later, Quintanilla felt obliged to share the Spaur file with Hynek, Hynek disagreed with the verdict, noting that the officers had clearly seen the object *in relation to* both the moon and Venus. He marked the case "unidentified," but it meant nothing. The case was "explained" and Deputy Spaur was left a sorry victim of the delusion hypothesis.

"Largely because the press and Blue Book concentrated on Dale Spaur almost to the exclusion of the other three witnesses, the public gained the impression that here was a case of one policeman's having become unbalanced and having experienced a major hallucination . . ." Hynek wrote. "Subsequently, Spaur was singled out for unbearable ridicule and the pressure of unfavorable publicity. The combination of events wrecked his home life, estranged him from his wife, and ruined his career and his health. He is no longer with the police force, and, it is reported, he subsists by doing odd jobs."[21]

Quintanilla himself summed up Spaur's story in his own brusque way: "This type of situation is extremely ticklish, because you always have a loser. It's a shame that it has to be this way, but it happens to be a fact of life."[22]

Although it proved to be a popular bestseller, Hynek's book also impressed a great many of his scientific brethren, and that was, after all, one of the key demographics he'd hoped to win over. "Hynek's *The UFO Experience: A Scientific Enquiry* (1972) clearly stands apart from earlier books written about UFOs," said one academic.[23] "On balance, Hynek's defense of UFOs as a valid, if speculative, scientific topic is more credible than Condon's attempt to mock them out of existence," wrote the reviewer from *Science* magazine.[24]

"Thank you for your book *The UFO Experience*," Claude Poher wrote to Hynek from the French space agency after Hynek had sent him a copy. "This is, in my opinion, the most important book written on the subject. I am sure it will change the mind of several scientists who are not informed on the subject.

"I think you have found the best way to attack the fortress of indifference," he continued. "The next important step is to be sure this book is read by those who are in charge of budget and planning for science.

"As for me, I am doing my best in order to inform them of the existence of this book."

Hynek was suddenly marketable, as he revealed in a letter to Jennie Zeidman. "We are going on the Eclipse Cruise to Africa—Mimi and I get free passage in return for some shipboard lectures on Astronomy and UFOs!!! Fellow lecturers will include Isaac Asimov, Neil Armstrong, and [science writer] Walter Sullivan!

"We are taking the boys," he added. "They'll be 10 and 11 then, and should be ripe for adventure. (I'm always ripe)."

The Eclipse Cruise turned out to be a once-in-a-lifetime experience for the family. Son Paul remembered the cruise as "one of the coolest outings we ever did." Not only did they visit Africa and witness a total solar eclipse, but they also got to sit at what must have been the most popular table in the ship's dining room.

"If there [was] only one thing that can make it better," Paul recalled, "it was that our family and one other family would share all our meals together at a table for two weeks, and that was Neil Armstrong's . . . Here I am, an eleven-year-old astronomer's son, four years after the moon launch, and I'm hobnobbing with Mr. Armstrong and his kids for two weeks. It was phenomenal."

Fittingly, Paul remembered that the eclipse itself, the whole reason for the adventure, more than lived up to its billing: "It was perfect," he said, "just spectacular." And Hynek's record with solar eclipses was evened out at 2-2.

"BOIANAI IS A VERY FORSAKEN PLACE," Hynek recalled of his long-anticipated 1973 journey to the community in Papua New Guinea, where Father William Gill had waved to UFO occupants in 1959 and they had waved back.[25] When he finally got the chance to visit the island nation tucked between Oceania and Melanesia in the South Pacific in order to investigate the Gill Close Encounter of the Third Kind, he found the traveling exhausting but the trip fully worthwhile.

When he arrived at his remote destination, Hynek met up with the Reverend R. G. Crutwell, who had originally documented the Gill case in 1959 and brought it to the attention of Australian UFO groups. With Crutwell serving as interpreter, Hynek interviewed six of the original witnesses and found that they recalled the events of those nights fourteen years earlier in vivid detail. "I'm going there to reconstruct the crime," Hynek had mentioned in an interview in the Northwestern campus newspaper before his trip,[26] and that he did. Audiotapes of the interviews reveal a sometimes confusing effort to get the facts straight between an interviewer and interviewee with very different frames of reference, and with a somewhat frustrated interpreter trying to bridge the void. When Hynek, for example, asked one witness if the UFO looked like a "fireball," the word seemed to have no meaning to the interviewee, so Crutwell had to break in and reframe the exchange in a way that made sense to both Hynek and the interviewee. At another point, Hynek seemed confused that a witness did not appear to distinguish between "evening" and "night." In the end, though, the witnesses' individual descriptions and overall consistency were convincing to Hynek. "From the facial expressions and gestures of the natives, I sensed that the event had been real as far as they were concerned," he said.[27]

After the stay in Boianai, Hynek visited with Father Gill in Melbourne and found the minister to be "a painstaking, methodical, unexcitable person—just the sort to stand calmly by and take notes at the height of the exciting action."[28]

"The dynamic between Hynek and Gill was essentially about establishing the credibility of the sighting," said Bill Chalker of the meeting in Melbourne. "Hynek, being an astronomer, it was easy for him, once he got those details and once he was on-site at Boianai and met some of the native witnesses. All that allowed Hynek to confirm to his own satisfaction that the Gill case was really a strong case and that the key object—the craft, and the beings that they waved to—certainly wasn't an astronomical thing."

Hynek had a lot to think about as he returned home to the United States. The Gill UFOs seemed to him to have been intelligently controlled, and he was still flummoxed by the "blue sparkling haze or aura, which was not connected directly to the object but separated by about a foot."[29] And then there were those entities . . . Even if Hynek could account for their existence, how could he account for their exasperating behavior? "Why did they just wave casually to Father Gill?" he later wondered. "Perhaps they are trying to tell us something symbolically; on the other hand, if they are so smart, why don't they tell it to us directly? Why didn't they land in Boianai? Maybe it isn't transmittable in terms of language. They seem to give you a little bit of an idea and then give you reasons against it or give you something contrary, to change your belief structure."[30]

"The case," he admitted, "has always intrigued me."[31]

HYNEK'S TRAVELS WERE WELL KNOWN AT NORTHWESTERN, and it took little time for the university to schedule speaking engagements for its star professor. A generous write-up in the September 21, 1973, *Daily Northwestern* reported that Hynek had delivered "his popular 'UFO lecture' . . . before a packed Technical Institute Auditorium."[32]

In front of an appreciative audience on his home turf, Hynek approached the topic of UFOs with a refreshingly casual style and a knowing sense of humor, as he shared a series of slides of both authentic and hoax UFOs that had been submitted to him over the years. According to the reporter, the images looked like anything from "a blurred picture of a banana split" to an obvious double exposure to a natural cloud formation. Just for laughs, Hynek threw in a slide of a sensational tabloid headline claiming "I Was Seduced by a UFO!"[33]

Then, recounting his recent adventure in Australia and Papua New Guinea, Hynek explained that witnesses such as Father Gill and his associates convinced him that something real had happened to them. That, he said, was why he titled his book *The UFO Experience* "instead of something that would sell more copies, like *The Sensuous UFO*," a sly play on the name of the bestselling 1969 sex manual, *The Sensuous Woman*.[34]

What Hynek didn't quite realize that day was that the United States was in the opening stages of what would become its first full-blown UFO "flap" in several years. Unidentified flying objects were being sighted in historic—and alarming—numbers all around the country, and Hynek's calm demeanor and thoughtful, reassuring approach to the phenomenon would soon be in demand like never before.

Memorable flaps, or waves, of UFO sightings had occurred in the United States before, most notably in 1896–97, 1947, 1950, 1952, 1957, and 1965–67, and major flaps had occurred in France, Italy, Great Britain, Norway, Canada, Spain, South Africa, Malaysia, Argentina, Brazil, Mexico, Australia, the People's Republic of China, the Soviet Union, and other countries as well. The surge of sightings in the United States that began in the summer of 1973 rivalled them all, and Hynek was not alone in wondering whether something portentous was occurring.

"By September the South seemed to be involved in a wave, and by October it was clear that the entire nation was in its grips," wrote historian David Jacobs.[35] And although Jacobs wrote that this wave "mirrored" previous waves, that was not entirely true. The 1973–74 flap was the first to take place in a post–Blue Book world, and as Jacobs himself pointed out, people now felt differently about UFOs.

"During the earlier sighting waves, television news had concentrated on giving vent to contactee claims or ridiculing legitimate UFO reports as part of a national 'silly season,'" he observed. "But in 1973–74, for the first time television news squarely confronted the UFO problem. CBS, NBC, and ABC gave the UFO sightings the fairest and most impartial coverage the networks had ever given the subject."[36] For the first time ever, Jacobs wrote, TV news reports "refrained from tongue-in-cheek humor, 'silly season' editorializing, or ridiculing witnesses."

That was, until late October, when the world's credulity would be put to the test yet again, this time by two chilling Close Encounter cases in the American heartland.

CHAPTER 18

THE SPUR

IT WAS THE DOOR THAT DID IT. When it opened, everything fell apart.

Up until that moment, Calvin Parker and Charlie Hickson still thought they might be able to understand what was happening to them, might be able to keep their grip on reality. It was such an ordinary October night, after all: a few relaxing hours fishing on the pier, then home to bed for a solid night's sleep before starting the day shift the next morning at F. B. Walker & Sons shipyard.

Charlie felt obligated to give Calvin some normalcy. Older by some twenty-three years than his fishing companion, he had promised his old friend, Calvin's dad, that he would help the boy adjust to his new job and new surroundings. Pascagoula, Mississippi, was vastly bigger than Calvin Parker Jr.'s hometown of Tuckers Crossing, and the city made the nineteen-year-old feel lonely and unmoored.

So they went fishing the night of October 11. The two men sat on the rusty pier, a few yards apart, watching the dark water and occasionally wondering out loud where the fish had all gone to that night.

By nine P.M., the night had settled in and the talk had gone quiet, and the only sound was the lapping of the river. But then there was another sound. At first the humming didn't even register. But in a second or two it had grown loud enough—or close enough—for both men to hear it.

Charlie reached back for fresh bait, and when he turned he saw the shape that went with the sound. Two flashes of intense blue light startled them, and Calvin, too, looked around to see what was approaching.

The glowing, humming object approached them from across the bayou, dropping from the sky and then hovering a few feet off the water. It was oval, like an egg, and featureless, perhaps thirty to forty feet across and ten feet tall. It gave off a pulsating blue light, the same hue as the flashes that had announced its presence. And it floated there, fifty yards away from them, while something inside prepared to come out.

"I looked to each side. I wanted to run," Calvin recalled years later in a video interview,[1] but he and Charlie were penned in; an auto salvage yard blocked escape on one side and the river blocked the other.

When the door opened in the side of the egg, Charlie and Calvin knew they should have taken their chances in the river. Three things—three creatures—floated silently out the door and came toward them over the water. In the blue glow the creatures seemed to be gray, and although they never touched the ground they had legs and were humanoid, but only just: they had torsos, heads, arms, and legs, but nothing else about them was right. Instead of hands they appeared to have lobster claws. The feet were more like rounded stubs. Pointed cones took the place of ears and noses. The heads and necks were fused into solid units and rested implacably between the shoulders. There were no eyes in the gray faces, and the mouths were small slits that never moved.

Nothing was said. One of the creatures made buzzing sounds, but they did not speak. They simply took both men by the arms—two of them holding Charlie and one supporting Calvin—and floated them back to that door.

"I was scared to *death*," Charlie recounted hours later in testimony to the Jackson County sheriff. "And me with a spinnin' reel out there—it's all I had . . . Calvin done went hysterical on me."[2] Calvin blacked out the instant he was lifted off the ground; it was the only way he could cope. He did not consciously know that he had entered the egg until much later, and that was for the best.

Like those of many UFO witnesses, Charlie's thoughts flashed ahead to the outcome of the encounter and he feared the worst: Would the beings kill him? Would they take him away somewhere? Would he ever see his loved ones again? "I kept thinking, 'They will dredge the river and with no bodies they will assume we have drowned and washed out to sea,'" Charlie recounted.[3]

Once through the door, Charlie was unaware of Calvin's whereabouts. He was conscious only of where he had been taken: a bright room with a spherical floating "eye" that emerged from the wall. Still weightless, but his body now rigid, Charlie felt the creatures leaning him back. The eye approached and seemed to peer at him.

"The 'eye' came closer and stopped about six inches from my face," he said. "I tried again to close my eyes, but some force kept them open. The eye lingered there for a while then started to move down my body and returned to move over my entire body. No pain, no sensation."[4]

As the examination went on Charlie lost sight of the creatures. The buzzing stopped and Charlie knew he was alone, and that was almost worse. Had the creatures gone to examine Calvin? Would they come back? They communicated nothing: no purpose, no intention, no concern . . . Charlie saw Calvin outside the object twenty minutes later, when the creatures floated Charlie back to the pier and deposited him on solid matter at last. The moment his feet touched the rocky ground, Charlie's legs buckled, and he fell to his hands and knees. He looked up to see Calvin standing a few feet away, frozen in place, with his arms outstretched. "The only thing I remember is that kid, Calvin, just standing there," Charlie said. "I've never seen that sort of fear on a man's face as I saw on Calvin's."[5]

Charlie started to crawl across the rocks to Calvin, but he could hear that buzzing again and knew they still weren't alone.

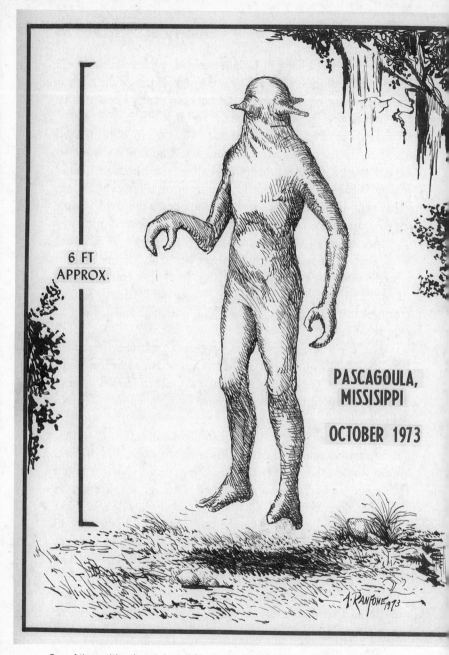

6 FT
APPROX.

PASCAGOULA,
MISSISIPPI

OCTOBER 1973

A. RANFONE, 1973

One of the entities that abducted Charles Hickson and Calvin Parker Jr. from a fishing pier in Pascagoula, Mississippi, and examined them aboard their glowing blue "spaceship." *(Artwork by Anthony Ranfone. Courtesy of Frances Ranfone Allain.)*

He rose, willing his legs to work, and turned to see what had become of the creatures. They were floating back to the craft and seemed to have no more interest in the two men. Charlie and Calvin watched them disappear into the brilliant interior of the egg, then the door closed and it was over. The egg made a buzzing sound and vanished.

Shock set in quickly. "I shook him, I even had to slap him, to get him to where I could talk to him," Charlie said.[6] After that he was able to calm the boy down enough to talk about what had happened, and it was a relief to both that they agreed on every detail. Calvin remembered nothing from inside the egg, but in every other aspect his experience mirrored Charlie's.

But what came next? What *should* come next?

It was after ten o'clock now, and the two men were back at their cars. Charlie helped himself to the whiskey he kept under his front seat as they considered their next step. "I asked [Charlie], I begged him, I said, 'Please, iet's not tell anyone, because I really don't want to look foolish. I don't want anyone to know about this,'" Calvin said.[7] Charlie felt the same way, but something nagged at him. What if that egg—and those creatures—were still out there? What if someone else was on their list for that night? What if there was a greater danger to the country or the world?

There was a phone booth outside a convenience store not far from their fishing spot, so once they calmed down somewhat they made their way to the Li'l General Curb Market and called Keesler Air Force Base in nearby Biloxi. God knows how their story came out, or what the receptionist on the other end of the call thought of it, but it went no further. The woman at the base knew the drill. As if reading from a cue card taped to her switchboard, she explained that the air force was out of the UFO business and instructed Charlie to notify the local sheriff.

"[Sheriff] Fred Diamond sent two detectives out to pick us up," Calvin said, "and as these detectives drove up I looked, and I could see them kind of laughing in the car, and I thought right then, 'This is the wrong thing, we shouldn't have done this, I know what's going to happen.'

"But when they seen the fear that was in our face, and when they talked to us, they knew right then that we were serious about it."[8]

Jackson County sheriff Fred Diamond heard them out and, to their relief, did not make light of their story. Perhaps it was because Charlie and Calvin both requested a lie detector test the moment they arrived at the sheriff's department, or perhaps Diamond knew that several people in the area had reported a strange object in the sky the night before.

As a matter of course, the deputies separated the men and questioned them about their experience. Their stories meshed, but Diamond and his deputies couldn't help but remain skeptical. After all, Charlie had whiskey on his breath, and what else could it be, really, but an unusually intense hallucination?

While Diamond thought it over, he reunited the two men and left them alone in an interrogation room. What Calvin and Charlie didn't know was that the sheriff had left a tape recorder running in the room, and it captured every word of their conversation. Diamond had expected the two to drop the act the moment they were alone and congratulate each other on pulling off such a brilliant hoax. What he got instead was one of the most chilling and unsettling dialogues ever recorded . . .

CALVIN: I got to get home and get to bed or get some nerve pills or see the doctor or something. I can't stand it. I'm about to go half crazy.

CHARLIE: I tell you, when we're through, I'll get you something to settle you down so you can get some damn sleep.

CALVIN: I can't sleep yet like it is. I'm just damn near crazy.

CHARLIE: Well, Calvin, when they brought you out—when they brought me out of that thing, goddamn it I like to never in hell got you straightened out.

CALVIN: My damn arms, my arms, I remember they just froze up and I couldn't move. Just like I stepped on a damn rattlesnake.

CHARLIE: They didn't do me that way.

(Charlie sighs)

CALVIN: I passed out. I expect I never passed out in my whole life.

CHARLIE: I've never seen nothin' like that before in my life. You can't make people believe—

CALVIN: I don't want to keep sittin' here. I want to see a doctor—

CHARLIE: They better wake up and start believin' . . . they better start believin'.

CALVIN: You see how that damn door come right up?

CHARLIE: I don't know how it opened, son. I don't know.

CALVIN: It just laid up and just like that those son' bitches—just like that they come out.

CHARLIE: I know. You can't believe it. You can't make people believe it—

CALVIN: I paralyzed right then. I couldn't move—

CHARLIE: They won't believe it. They gonna believe it one of these days. Might be too late. I knew all along they was people from other worlds up there. I knew all along. I never thought it would happen to me.

CALVIN: You know yourself I don't drink.

CHARLIE: I know that, son. When I get to the house I'm gonna get me another drink, make me sleep. Look, what we sittin' around for. I gotta go tell Blanche . . . what we waitin' for?

(Calvin grows panicky)

CALVIN: I gotta go to the house. I'm gettin' sick. I gotta get out of here.

(Charlie gets up and exits, leaving Calvin alone.)

CALVIN: It's hard to believe . . . Oh God, it's awful . . . I know there's a God up there . . .⁹

Before he went home, and still unaware of the secret recording, Charlie had a talk with Sheriff Diamond. "I made him promise me that there'd be no publicity about it," he said. "So we left there with that assurance."[10]

By the time Charlie got home to his wife, Blanche, it was almost sunup, and he had to go to work shortly. He took his responsibilities as foreman at the shipyard seriously and wasn't about to let his traumatic night keep him from his job. As it turned out Calvin felt the same way and came in to work as well, ready to get everything back to normal. Shortly after getting Calvin and the rest of his crew to work, however, Charlie got a phone call in his office from a reporter in Jackson. "He was asking me what had happened to me and Calvin the night before," Charlie said. "Well, I just hung up on him. How did they find that out?"[11]

The answer came soon enough. The phone rang again and Charlie heard Sheriff Diamond's voice on the other end. "I said some bad words to him, and told him that news reporters had already been trying to contact me," Charlie recalled. "He said, 'I don't know how, Charlie, but it got slipped out somewhere.'

"And in that day, it went all over the world."[12]

For the second time in less than twenty-four hours, Charlie and Calvin found themselves hemmed in, with no escape route. Reporters started to show up at the shipyard, and the two men were summoned to the general manager's office to describe their abduction to their bewildered boss.

"Johnny Walker, the owner of the shipyard, had come in by this time," Charlie said. "After telling them our story, we discussed it with them quite thoroughly. Johnny mentioned that we should have legal counsel in a situation like this. We agreed, and so Joe Colingo, a well-known attorney in Pascagoula, was called."[13] Perhaps Charlie misunderstood his boss's intentions, for Colingo was the attorney for the Walker shipyard and was summoned initially to help Walker keep the hordes of reporters from interfering with the operations of the company.

"Johnny Walker called me up and said, 'You've got to come down here and help me figure out how to keep these reporters from getting in the way,'" Colingo said. "It was just causing havoc down there." Colingo explained that he was never called on to represent Charlie or Calvin legally, and he never charged the two men for his services. "*I* didn't think they were crazy," he added, "but I think some people thought I was crazy for not thinking *they* were crazy."

Colingo's first priority was to get Charlie and Calvin—and the press corps—off the premises. Better to have the reporters jamming the parking lot at the sheriff's office than blocking the entry gate to the shipyard. The second priority, for Walker's sake as much as for Charlie's and Calvin's, was to establish the men's credibility and remove any doubt that this was a hoax or a practical joke. So it was that when Colingo and his wards arrived at the sheriff's office, he had a request for Fred Diamond: Could the two men take that lie detector test?

In New York City, NBC News producer Fred Freed watched the reports from Pascagoula closely. Freed, an Emmy and Peabody award winner, had recently wrapped a prime-time documentary on the energy crisis and was already at least a month into preproduction on a prime-time special about the UFO mystery when news of Calvin Parker and Charlie Hickson hit the wires.

The UFO field was difficult, unfamiliar territory for any journalist with a reputation for presenting an audience with the unvarnished truth, and Freed was taking a risk. He was working closely with NBC reporter Ralph Blum on the documentary, and the two men had already retained Dr. Hynek as a consultant on the project, but still . . . short of capturing a flying saucer landing outside Rockefeller Center, how could Freed prove anything one way or another?

"In October 1973, NBC had arranged for Dr. J. Allen Hynek to come to New York to consult about a documentary program on UFO's," Blum reported the following year in the *National Enquirer*. "Dr. Hynek was already on his way to New York on October 12 when the wire services carried the story that a UFO had landed in Pascagoula, Miss., and two men had been 'floated' aboard by 'creatures.' Twenty-four hours later, I found myself on a plane headed for Mississippi accompanied by Dr. Hynek, who had cancelled all his plans because Pascagoula sounded like an authentic UFO encounter."[14]

However, Hynek admitted in a radio interview months later, "I went down to Pascagoula . . . completely negative."[15] Indeed, in *The UFO Experience,* he had admitted to a serious ambivalence regarding incidents such as the one he was going to investigate in Mississippi.

"We come now to the most bizarre and seemingly incredible aspect of the entire UFO phenomenon," he wrote. "To be frank, I would gladly omit this part if I could without offense to scientific integrity: Close Encounters of the Third Kind, those in which the presence of animated creatures is reported . . .

"Unfortunately, one may not omit data simply because they may not be to one's liking or in line with one's preconceived notions."[16]

In later years, he confessed to having undergone an epiphany in Mississippi: "My thinking was altered completely when I was called in . . . to interrogate two Mississippi fishermen . . . who insist they were literally 'kidnapped' and forced to go on board a spacecraft, where they were subjected—just as in the case of the Hills—to a physical examination," he said in an interview. "The tale told by these rugged shipyard workers held up under grueling cross-examination."[17]

Hynek confessed that the tape recording of Hickson and Parker made in secret by the sheriff was the clincher. He was struck by the fact that the two men stuck to their story even when they assumed no one could hear what they were saying. "I don't know what makes me want to automatically look down upon these creature cases," Hynek went on. "Maybe this involves an atavistic fear of the unknown, or of rivalry with another species. There is, upon closer scrutiny, another factor which I find difficult to sort out. It is odd that the creatures seen coming from these craft should resemble our own *homo sapien* race so closely."[18]

Perhaps Hynek's time with Father Gill a few weeks earlier had softened his doubts about alien encounters and made him more

receptive to Hickson's and Parker's story. And now he was stepping into what was quickly becoming a media circus in the heart
of the Deep South. Colingo recalled that Hynek's arrival at the
Walker shipyard actually brought a measure of calm to the frenzied proceedings. "Hynek was introduced to me as someone who
had worked on Project Blue Book for the air force and who had
proper credentials to know what he was talking about."

When Hynek arrived on the scene, Hickson and Parker had
already recounted their story at least six times: twice to Sheriff Diamond and his deputies (not counting the secret recording), once
to their employer, once to Joe Colingo, once to UFO researcher
Dr. James Harder, and once to security personnel at Keesler Air
Force Base. In every retelling, the story remained entirely consistent. The details never changed.

One might wonder how Hickson and Parker could have been
ushered into Keesler AFB after being turned away the night before. By the time Colingo returned the men to the sheriff's office Friday morning, the abduction narrative had taken hold and
people were genuinely wondering if something unearthly had visited Pascagoula the night before.

"Nobody knew what had really happened, except for Hickson
and Parker," Colingo said. "The concern was: If they were picked
up by some kind of foreign spacecraft, could they have radiation
connected to them? They wanted to know if they were radioactive." It seemed prudent to find out, but the only hospital in
the area with the facilities to run such tests was the infirmary
at Keesler AFB. Because the request for testing came from the
Jackson County Sheriff's Department, the air force had at least a
semblance of cover for getting its hands dirty with a UFO case,
so later that morning they brought Hickson and Parker, accompanied by Joe Colingo and Detective T.E. Huntley, to the base
infirmary.

The radiation tests came back negative, but because the two
men were never officially checked in to the infirmary, there was
no record of the actual tests given or of the actual results. Just
the same, the air force's collective curiosity had, apparently, been

piqued. So it was that Hickson and Parker found themselves seated across a table from a military intelligence officer and the base's top brass, telling their tale once again.

The air force's reaction to the interview was simple and predictable: there was no demonstrable threat to national security, so there would be no investigation. When reporter Murphy Givens of the *Mississippi Press* inquired how the air force planned to proceed with the Pascagoula case, he was told the air force would contact the staff members of the astronomy department at Tulane University and encourage them to investigate the incident.

Hynek finally met Hickson and Parker at the Walker shipyard offices on Saturday, two days after their experience, and found them exhausted and overwhelmed, but well prepared to go over the details of their experience one more time.

James Harder, an engineering professor at Berkeley and consultant to APRO, had arrived from California earlier in the day and had already interviewed the men by the time Hynek arrived. Harder was ready to hypnotize them, but there was a problem: Hickson and Parker refused to be hypnotized. They didn't understand it and they didn't trust it.

A frustrated Harder decided that a local psychiatrist might be able to help convince Hickson and Parker that hypnotism was safe. Like lawyers, psychiatrists were in short supply in Pascagoula, and there seems to have been only one man to call: Dr. William D. Bridges. He joined Hynek, Harder, Parker, Hickson, and several of the sheriff's deputies in an inner office, and there they remained for hours.

Later in the day, Hynek told Ralph Blum that even Dr. Bridges's presence wasn't helping. Harder had been hypnotizing other people in the room, so that Hickson and Parker could see for themselves that it was safe, but the two men weren't even close to convinced. Blum then asked Hynek what he thought of Hickson and Parker's tale. "It's one of the most amazing I've ever heard," Hynek replied after some thought. "I think they're telling the truth. I think they had a genuine UFO experience."[19]

Saturday afternoon, when Hynek and Harder held a press conference, it became obvious that the hypnosis session had had just as profound an impact on the scientists as it had on the witnesses. The answers for which everyone had been waiting throughout that long afternoon never materialized, however, for the hypnosis attempt had failed spectacularly.

Eventually Hickson and Parker had consented to hypnosis and Harder was able to put them under. Both men had panicked when asked to relive their experience on the pier, however, and on Bridges's insistence were brought out of hypnosis to prevent any psychic damage from occurring.

In the end, fear settled all doubts. Hynek and Harder never got the interview they had hoped for, but they both found Hickson's and Parker's panic and terror to be genuine and convincing. "There's simply no question in my mind that these men have had a very real, frightening experience, the physical nature of which I'm not sure about," Hynek told the press, "and I don't think we have any answers to that. But I think we should very definitely point out that under no circumstances should these men be ridiculed. They are absolutely honest. They have had a fantastic experience and also I believe it should be taken in context with experiences that others have had elsewhere in this country and in the world."[20]

Harder dove right into territory that Hynek had been careful to avoid, insisting that Parker and Hickson had come face-to-face with an extraterrestrial phenomenon. "I can say so beyond any reasonable doubt," he claimed. When the reporters asked whether UFOs posed any threat to the human race, Harder said, "If you pick up a mouse in a laboratory situation, it's very frightening to the mouse. But it doesn't mean that you mean the mouse any harm."[21]

Hynek told the reporters that the incident was deserving of further investigation, even though he knew that the chances of that were meager. In his wrap-up of the press conference, the *Mississippi Press*'s Givens reported the following: "Such investigations, said Hynek Saturday night, usually take up to six months to complete. But, he said, such investigations take money and neither he nor Harder could pursue the matter on their own."[22]

Of course, at that moment Hynek wasn't on his own. He had Fred Freed's ear at NBC, and maybe that could be turned to his advantage.

"Experts Back Stories of 2 Who 'Saw a UFO'" said the headline in the October 15 *Chicago Daily News,* and similar headlines were appearing in newspapers around the country and around the world. The October 21 *New Orleans Times-Picayune* quoted Dr. Lew Epstein of the Louisiana State University Astronomy Department as saying, "Dr. Hynek is a very conservative, uptight square who would be the last person to take the bait of a flying saucer story. Hynek's conclusions lead me to believe it is possible something extraterrestrial did occur in Mississippi. At least it's worth a thorough investigation."[23]

Many would still need more convincing, but not Joe Colingo, who came out of those remarkable days believing that Hickson and Parker had far surpassed the threshold of credibility: "They were scared. They were not faking that at all. It didn't take a doctor or a genius to see that. They saw something; I don't know what it was, but they saw something that scared the fire out of them." For months after the incident, Hickson would occasionally call Colingo for advice on how to respond to media requests or interview questions, and Colingo never once detected a false note. "Charlie Hickson never gave me any reason to think he was lying."

Givens noted in the *Mississippi Press* that "the two men now pass a small leaflet, printed: 'We do not want to talk to anyone, or be interviewed by anyone, at this time, about our recent UFO experience. Our lawyer, Mr. Joe Colingo, will arrange a public interview during next week. Until then, please do not invade our privacy. We need to rest and also work.'"[24]

The day after the Pascagoula press conference, Hynek returned to Evanston, but he could not escape the press or the public. Every time Pascagoula came up in conversation or during an interview, Hynek was asked, "Did it really happen?" On October 18, only one week after the Pascagoula incident, Hynek addressed the local chapter of the Research Society of America at Argonne National

Laboratories outside of Chicago, and he addressed that very question in a thoughtful and nuanced way.

As the *Argonne News* reported, "[Hynek] made the point that UFOs should not be thought of as synonymous with visitors from outer space, though neither do they deny such a possibility. Hynek said they may really be a signal for '. . . tapping a new domain in nature. We may be on the border of something which we truly can't yet understand any more than Plato could have understood television.' He said, 'It is the thing that doesn't fit, that rankles, that to the scientist is the spur to further action.'"[25]

A FEW HOURS AFTER HYNEK addressed his colleagues at Argonne Labs on the eighteenth, the spur struck again.

At about eleven o'clock that night, Army Reserve Captain Lawrence Coyne and his crew—First Lieutenant Arrigo Jezzi, Sergeant John Healey, and Specialist Second Class Robert Yanacsek—were flying their UH-1 Huey helicopter ambulance on a routine flight from Columbus, Ohio, back to their home base in Cleveland. The four men had spent the day in Cleveland for their regular medical exams and were heading home with clean bills of health. It was a routine flight along a familiar route, and skies were clear. Jezzi was in the pilot's seat, and he was keeping the chopper steady.

"SSG Yanacsek observed a red light on the east horizon, 90 degrees to the flight path of the helicopter," read the crew's report to their commanding officer. "Approximately 30 seconds later, SSG Yanacsek indicated the object was converging on the helicopter at the same altitude at an airspeed in excess of 600 knots and on a midair collision heading."[26]

As the light grew threateningly close, Coyne took over the controls and put the chopper into a "power descent from 2500 feet to 1700 hundred feet."[27] He checked in with the nearest base to ask if there were any experimental aircraft in the area, but he got no reply. Meanwhile, the red light grew brighter and closer, and "the crew expected impact."[28]

Coyne put the helicopter into a precarious dive that would

bring them close to the treetops in seconds. It was a dangerous move, but he had no alternatives. "It looked like a torpedo or missile locked onto us," Coyne said.[29]

Then, suddenly, the red spot was something much more. A metallic, cigar-shaped object, some fifty or sixty feet long, now filled the chopper's windscreen. The red light was on the left, on the craft's apparent "nose," while a white light glowed at the object's rear. A green light beamed from the aft end of the craft and bathed the cockpit of the helicopter in its light.

There was nothing familiar about the craft: "no wings, no landing gear, and no vertical or no horizontal stabilizers. A dome-like affair 12–15 feet in diameter was on top of the craft. Its shape blotted out the bright nighttime stars."[30]

The crewmen in the helicopter shared a sensation that the object "hesitated" and remained motionless ahead and above of the chopper for ten endless seconds. When the strange craft streaked away to the northwest and the white light at its tail was lost to their view, Coyne made an astonishing discovery: despite the fact that he was still holding the controls in a dive orientation, the altimeter showed the helicopter was now at an altitude of 3,500 feet and still climbing at 1,000 feet per minute.

"A Huey can't do that," Coyne related.[31]

Moments later Coyne regained full control of the helicopter, and the four men resumed their flight to Cleveland without any further sightings. The event had lasted about two minutes, and within ten minutes the helicopter's radio began working again.

It would be three days before the sighting was reported in the *Cleveland Plain Dealer*. The Coyne incident quickly became national news and, along with the Pascagoula case, capped the 1973 flap in spectacular style.

"By any standard," Jerome Clark wrote of the Coyne Close Encounter, "it is one of the most important UFO events ever recorded, in no small part because it received the kind of thorough investigation it merited. Never satisfactorily explained, extraordinary in its implications, it stands as a challenge to those who would reduce all of the UFO phenomenon to error, delusion, or deceit."[32]

CHAPTER 19

PURPLE PEACH TREES

TWO WEEKS AFTER THE HICKSON–PARKER abduction and one week after the Coyne incident, Hynek was storming the gates at Wright-Patterson Air Force Base, demanding that his employers look into the rash of UFO sightings sweeping across the nation.

He had already planned a trip to Ohio that week, to accept an Ohioana Book Award from the Ohioana Library Association for *The UFO Experience,* and Jennie Zeidman was to attend the award luncheon with him on October 27. In light of the Coyne event making national news on the twenty-first, Hynek informed Zeidman that he would be coming to Ohio a day early, and he asked his old friend if she could pick him up at Wright-Patterson in Dayton at the end of the day and drive him to Columbus.

This was not an unusual request. After all, as a part of Hynek's mysterious post–Blue Book consulting contract, he had been making periodic visits to Wright-Patterson and maintaining ties with colleagues in the Foreign Technology Division. This was, however, not a good time to be making an unscheduled visit to an air force base. Egypt and Syria had attacked Israel only a few weeks earlier, and U.S. military bases were on high alert. Nonetheless, Hynek returned to the former home of Project Blue Book to make his case to whoever would listen.

When Zeidman rolled up to the gate at the base at the end of the day, she found Hynek in a state of distress. What Zeidman learned was that Hynek's request for air force backing of an investigation into the current UFO flap was turned down cold. "One of the most intense UFO flaps was in full swing, and the Air Force was doing nothing," she recalled. "[Hynek] had challenged them. 'Why aren't you doing something?' he later told me he had said. And he had received only stony faces in reply. It wasn't a question of a Mid-East-War-going-on-and-we-can't-be-bothered-with-UFOs. It was a question of, regardless, the Air Force 'would do nothing about' UFOs."[1]

"Allen thought that it was plausible that the air force would, at least, support a small-scale effort," said Mark Rodeghier. "This was a case of Allen being Allen, meaning still hopeful and naive after his long, mostly negative association with the air force."

BACK IN MISSISSIPPI, the Hickson-Parker case continued to unfold in its own haphazard way.

It wasn't until October 30 that Joe Colingo's attempts to arrange a polygraph test bore fruit, but on that day only Charlie Hickson was available. In the three weeks that had passed since the incident, Calvin Parker had been, in his words, "torn apart." It's entirely possible that Dr. Harder's attempt to hypnotize Parker on the thirteenth had resulted in "psychic damage," as Dr. Bridges, the hometown psychiatrist, had warned. All that is known is that hours before he was to undergo his lie detector test, Parker suffered a breakdown and was admitted to a psychiatric hospital, leaving Hickson to shoulder the burden of public scrutiny on his own.

"We called a company out of New Orleans to come to Pascagoula, and they gave Hickson a polygraph," Colingo recalled. "I wasn't in the room, but I can still remember the quote that the polygraph operator said. He said, 'I'm not saying he saw a spaceship, but when he *said* he saw a spaceship, he wasn't lying.' That's the way he put it."

Meanwhile, Hynek's investigation had uncovered some jarring details about Parker and Hickson. Apparently, Parker had been known to tell tall tales when he was a schoolboy, tales about going hunting for "giant rabbits" and other absurdities. Hickson's history was even more troublesome: he had allegedly been the getaway driver in a botched 1959 robbery and had served jail time. There was a story that he had been fired from a job once for excessive drinking, and according to an unnamed niece, "Charlie is prone to imbibe quite a bit" and had "seen things" like the Pascagoula UFO before.

There was something else about Hickson. He never wore a watch. Watches never kept time once he slipped them onto his wrist, he claimed, so he just gave up wearing them. No one could ever explain it, and no one seems ever to have tried.

Despite the events in Mississippi and Ohio, NBC News scrapped Freed's UFO special. "There was just too much happening in the world," said authors Ralph and Judy Blum.[2] The Watergate scandal was heating up, the energy crisis was affecting the daily lives of millions of Americans, and war was still tearing up the Middle East; not even UFOs could compete with such a barrage of crises.

But UFOs were still news, and from the ashes of the NBC special arose invitations for Hynek and Hickson to come to New York City to appear—along with Captain Lawrence Coyne and noted skeptic Carl Sagan—in prime time on *The Dick Cavett Show* on rival TV network ABC. Cavett, a comedian-turned-talk show host, generally interviewed celebrities on his show, so it was unusual for him to focus on such a newsy topic. But he and his producers couldn't have missed seeing Charlie Hickson, Captain Coyne, and Dr. Hynek on TV over and over again through late October. Maybe they saw sincerity or maybe they saw ratings, but their entire November 7 episode was devoted entirely to the theme of contact with UFOs.

It is unfortunate that the company that owns the rights to *The Dick Cavett Show* apparently no longer possesses a copy of this particular episode, because it would be a real collector's item today. When asked about it, the director of digital media for Daphne Productions explained that "it was taped over back in the 70's."

Because of this oversight, we can only relive the melee through a partial transcript recorded by Ralph and Judy Blum in their 1974 book, *Beyond Earth: Man's Contact with UFOs.*

As the Blums described it, when the lights came up on the soundstage, Cavett introduced the show in dramatic fashion, holding up a sheet of white paper that he identified as the report on Charlie Hickson's polygraph test. Cavett told his audience that the report was written by Scott Glasgow, polygraph operator for Pendleton Detective Agency of New Orleans, Louisiana. He then read the money quote from Glasgow's report: "It is my opinion that Charles Hickson told the truth when he stated (1) that he believed he saw a spaceship, (2) that he was taken into the spaceship, (3) that he believed he saw three space creatures."[3]

Then Cavett introduced Hickson to two million viewers and proceeded to interview him carefully, thoughtfully, and respectfully for the next fifteen minutes.

In the green room, waiting for his segment of the show, Hynek could not have been happy. Spaceship? Space creatures? Now that Cavett had introduced the show with those damnable descriptions, and inadvertently set the tone for the next ninety minutes, how could he counter that?

Also present in the green room was Hynek's cordial antagonist, Dr. Sagan, who must have taken in Cavett's introduction with pure delight. Now that the evening's conversation was officially about flying saucers and little green men, Sagan, whom the Blums described as a "darkly handsome young man in a white suit,"[4] was free to ridicule at his discretion.

After Hickson's interview, Cavett talked with Captain Coyne, who told of his terrifying Close Encounter just a week after Hickson's experience. Next, Cavett introduced astronaut James McDivitt, who recounted his UFO sighting while in space on NASA's Gemini 4 mission. Then came author John Wallace Spencer to discuss his new book about the Bermuda Triangle.

When Hynek finally took the stage, he avoided mention of space creatures and spaceships and announced the founding of a new scientific research organization dedicated to the study of the

UFO phenomenon. The organization would be called the Center for UFO Studies (CUFOS) and would be headquartered in Evanston, perhaps on the Northwestern University campus itself.

The announcement may not have meant much to the vast majority of people watching the show, but for many—including Sagan—this was a moment of great significance. Not only was Hynek making a serious attempt to finally bring scientific rigor and credibility to UFO research, but he was also signaling that *he had support from others in the scientific world.*

Sagan wasn't in the mood for acquiescence and quickly "zapped" Hynek with what the Blums called "The Sagan Method." Portraying the allocation of research funds as a zero-sum game, Sagan declared that any resources that might be devoted to establishing a UFO research center would be tragically misplaced and misspent. He felt that his own planetary research was the "*best* use of resources."[5]

"One by one, Sagan disposed of the other guests by mocking their testimony," wrote the Blums, "and he did it with real flair." He compared Hickson with people who report seeing leprechauns. Then, when astronaut McDivitt came to Hickson's defense, Sagan doubled down and compared Hickson with people who believe they've made contact with deities, or who think they *are* deities. He attacked Spencer's book by asking why freight trains never disappear into the Bermuda Triangle. He dismissed Captain Coyne's sighting as a hallucination, despite the fact that there had been a crew of four aboard the chopper, all of whom experienced the same thing. And when Hynek cut in to defend Coyne by telling Sagan that "altimeters don't hallucinate," Sagan replied, "I don't mean to attack Captain Coyne but people who *read* altimeters hallucinate."[6] Sagan didn't so much argue his points as throw pies. Afterward, the Blums aptly defined Sagan as "a master of forensic gamesmanship."[7]

Hynek may have wished he had followed Jacques Vallee's lead and turned down the invitation to appear on the show. As Vallee wrote in his journal, "When I watched the Dick Cavett show I congratulated myself for not going to New York to take part in it. Allen looks indecisive, timid and tired while Carl Sagan takes control of the discussion from beginning to end and ridicules UFO research."[8]

Sagan was so glib and facile—the Blums described him as prefacing his statements by tossing his head and "throwing back a thatch of black hair"—that the audience never caught a glimpse of the chinks in his own armor. Far from disbelieving in intelligent life elsewhere in the universe, Sagan had for years been leading the field in attempting to locate and communicate with extraterrestrial life-forms. It's not that he didn't think life existed elsewhere, or that other civilizations could visit Earth. He simply believed that if those visits had occurred, they took place in the distant past, so there was no reason to think that aliens from other worlds were visiting Earth en masse in 1973. It was an argument the Blums dubbed, "Ancient, Yes! Modern, No!"[9]

In the end, Hynek quietly broke in with this showstopper: "Well, if these are intelligences, then they know something about the physical world that we don't know, and they also know something about the psychic world that we don't know—and they're using it all."[10]

Hynek, too, was using it all. The Pascagoula incident may have made him uncomfortable in many respects, but it helped pave the way for the Center for UFO Studies and reemphasized for Hynek the need to focus on the human element of a Close Encounter case. On the final page of his Pascagoula case notes, Hynek noted, "Parker started crying in sheriffs office. Hixon [sic] seemed to get emotional on Cavett show, like it happened."

It's hard enough to do a good job investigating one blockbuster case, much less two in one week. It would be several months before Hynek could devote any attention to the Coyne case, in large part because he was still struggling to understand the difficult data presented by Parker and Hickson. "I talked and worked with those men quite a while," he said in an interview. "I listened to tapes that were taken when they didn't know they were being taped, I saw how Charlie behaved under hypnosis, and finally the lie detector test. All of those things convinced me that he was not making it up. They had had an experience, period." When the interviewer asked Hynek if he could be certain of their story, he replied, "There's no way I know of which could determine that. It's like if you told me you dreamed of purple peach trees last night, what can I do about it?"[11]

LESS THAN TWO MONTHS after Hynek and Sagan squared off against each other on *The Dick Cavett Show,* the *Chicago Sun-Times* offered them a rematch. In a December 30, 1973, feature article, reporter Dennis V. Waite delved into the ideological abyss that separated the two distinguished scientists in an attempt to find out how they could agree on the existence of intelligent life elsewhere in the universe, yet disagree so vehemently on virtually everything else.

Waite immediately set up a visual contrast: Hynek was described as "a goateed, soft-spoken Chicagoan" who "narrows his eyes and becomes serious" over UFOs, as he "sips cocoa and nibbles on a sandwich," while Sagan, the New Yorker, was described as a "modish-looking" professor who "grins at the mention of UFOs" while "delicately munching a chocolate French pastry."[12]

One can only read "modish-looking" to mean one thing: turtleneck.

Waite recounted Sagan's success at having an inscribed gold plaque attached to the Pioneer 10 deep space probe, a cosmic message in a bottle that described humans and our planet to any alien lucky enough to find it. While Sagan deeply believed that efforts to contact other civilizations was a worthwhile endeavor, he poohpoohed any notion that those same life-forms would come visiting Earth, noting that human civilization is "nothing special."[13]

"I find that we ought to be visited rarely, perhaps every 100,000 years. Perhaps every million years," Sagan said. His vague reasoning was that a million years is "a characteristic time scale for developing new species."[14]

Hynek, meanwhile, was described by the reporter as "desperately searching for something scientists can handle" from his office at the Lindheimer Astrophysical Research Center.

"Hynek today does not claim UFOs are extraterrestrial space-craft," Waite wrote, "but, he emphasizes, science must stop giggling at the mystery. The Air Force's investigations and the University of Colorado's 18-month probe were both shams, he states flatly. 'What sort of scientific investigation is it,' Hynek snaps, 'that as-sumes the answer before starting?'"

Waite then went into some detail in describing Hynek's na-scent Center for UFO Studies, noting that "some 4,100 police chiefs across the nation have received information about the UFO Central. A toll-free telephone number is being provided them and to the military, commercial airports, radar installations, civil defense units, and others. Field investigators from various UFO associations will be sent out to research the more substantial sight-ings and encounters."

Notably, Hynek signaled that Close Encounter cases would be of special interest to CUFOS, as they were most likely to be of sci-entific value. Also notably, Waite mentioned that CUFOS could cost $100,000 a year to operate.

"Sagan . . . wishes Prof. Hynek well in his efforts," Waite re-ported, then added: "The New York astronomer would prefer, however, that the money be spent on space science."[15]

When Hynek had a chance to speak with Coyne, Yanacsek, and Healey about their encounter, he elicited an almost mystical reaction from Captain Coyne: "This thing was so . . . above my comprehension," Coyne said before trailing off, words failing him. Clearly, the attacks on Coyne's credibility had left marks. "Some-times, after I get done talking to these people, I begin to wonder about it myself," he said. "But I figure, the hell with it, I'll just tell what I saw and that's . . ."

"Well . . . remain honest," Hynek interjected. "That's all you can do."

When Hynek asked Coyne whence most of the criticism had come, Coyne said Sagan had been negative "from the beginning." "Dr. Sagan was the first man I talked to with a different . . . ap-proach," Coyne said politely. "And . . . like I said on the [Cavett] show, I would like to see *someone* come up and say, 'Well, here's what you saw.'"

When Hynek asked how he was regarded among his colleagues, Coyne said, "Oh, they razz you! They don't know. They all say they *wish* they could have seen it . . . But, when you sit down and say, 'Look: I hope someday *you* see the thing, and I want to see how *you* react,' then, right away they become very sober about it."

Yanacsek said that he also "took some flack" both in his guard unit and at his workplace. Then he defiantly uttered what has become a battle cry for UFO witnesses everywhere: *"I know what I saw."*

The other witnesses who eventually came forward felt the same way.

When Jennie Zeidman revisited the Coyne investigation a few years later, she found that local UFO investigators had discovered additional witnesses, a woman and four children, who had seen the entire event from the ground.

The witnesses told Zeidman that they were driving home from Mansfield, Ohio, that night when they saw red and green lights descending over the nearby treetops. At first the lights seemed to be approaching the car, but then the witnesses became aware of the clatter of the approaching helicopter and pulled over to observe. Two of the children got out of the car and watched as the strange object bathed the helicopter, the trees, even the road and their car, in its green light. The object seemed to hover over the helicopter for about ten seconds.

Moments later they caught sight of the object again as it performed several course changes, appearing to zigzag across the sky before taking off toward Mansfield. "A flight path," Zeidman reported, "which correlates perfectly the motion of the object established through analysis of the aircrew's report."[16]

Zeidman then located witnesses who had, on that same night, seen and heard something strange in the sky above their rural home, less than two miles from the site of the incident. The woman who lived in the house was terrified by the cacophony of what seemed to be a dangerously low-flying helicopter. She buried her head under her pillow in fright, only to hear her son shout out from his bedroom asking if she, too, could see the green light.

A GALLUP POLL taken in late 1973 was not good news for Sagan: "An astonishing 11 percent of the adult population, or more than 15 million Americans, say they have seen a UFO—double the percentage recorded in the previous survey on the subject in 1966," said the Gallup press release. Gallup went on to report that 51 percent of adult Americans believed that UFOs were real and not "figments of the imagination or hallucinations," and 46 percent believed that intelligent life existed on other planets.[17]

Newspapers and magazines, meanwhile, sensed the shift in the public's mood and began running serious, sympathetic stories about UFO research. Headlines such as "UFO Studies Center Gaining Support," "Flying Saucers: New Respect—Less Publicity," "UFOs Real to Many Americans," "People Don't Scoff So Much Anymore at UFO Researcher," and even the charming "Hey, Mr. Spaceman!" signaled, if not a full embrace, then a new acceptance of the phenomenon and what it meant to the public. NBC even reversed course at the end of 1974 and aired an hour-long documentary entitled *UFOs: Do You Believe?* that offered a surprisingly nuanced view of the phenomenon. UFO historian David Jacobs found the NBC show both encouraging and discouraging, which sounds suspiciously like a reaction to unexpected journalistic balance. "I was pleased that NBC gave the phenomenon what it considered to be a 'fair' presentation," he wrote to Hynek that December, "but I was disappointed that it felt that the program should contain 'balance' which usually means that you have . . . jerks like Sagan . . . to poo-poo the whole thing."[18]

Despite the "poo-pooing," Jacobs had to admit that NBC had done a creditable job. "The simple fact that the show was a sober, nonbullshit, nonridiculous discussion of the various UFO groups and investigations is the most important thing," he wrote. "One must remember that a television network has never before presented a show even vaguely receptive to the idea that UFOs are a serious phenomenon."[19]

In this new, UFO-friendly world, Hynek was a legitimate star, and he rarely declined an invitation to talk about the center. As one Northwestern alumnus wrote, "If Hynek isn't bionic, he certainly is ubiquitous." He even started showing up in gossip columns; Irv Kupcinet's famous Kup's Column, a longtime fixture in the *Chicago Sun-Times*, often found column space to spread the word about Hynek's activities. "Kup" also followed the comings and goings of the great jazz drummer Buddy Rich, who had seen UFOs himself and always found time to visit Hynek and "talk shop" when his tours brought him to Chicago. Kup even announced the formation of CUFOS in his December 27, 1973, column, boasting that the center would be "the only academic-oriented, nongovernment body involved in this study." And, to recognize the other budding celebrity in the family, Kup announced in his June 9, 1973, column that "Mrs. J. Allen Hynek, political activist and wife of famed Northwestern U astronomer, is running for alderwoman of Evanston's 7th Ward."

Sadly, even Kup couldn't help the cause: Mimi lost her election by 393 votes.

HYNEK LAUNCHED CUFOS WITH A BANG, boasting a scientific board comprising twenty-two colleagues from around the world—many of whom still preferred their involvement to be, for the time being, of the anonymous variety. There were some expected names—Hynek, Vallee, and Fred Beckman from the original Invisible College, of course—and some not-so-expected names: David Saunders, from the Condon Committee, was on the list. So was Claude Poher, the French space scientist Hynek had met at NASA. So, too, was Thornton Page.

The roster included physicists, astronomers, biologists, psychologists, an aeronautical engineer, a mathematician, a meteorologist, a plant physiologist, and a psychiatrist. "The Center," Hynek wrote, "provides an organization wherein these people may work in free association on this problem." Members would, accordingly, put in time when they could, and meetings would be held informally whenever two or more members happened to be in Chicago or in attendance at the same conference. "It is more of an organization than a geographical place," Hynek explained. "Any professor connected with the Center, regardless of which university he is affiliated with, is part of the Center.

"We have been offered the use of several laboratories and a significant number of scientists have shown an interest."[20]

This, Hynek felt, was how science should be done in the late twentieth century. His experiences at Johns Hopkins and White Sands had pointed the way toward a collaborative, multidisciplinary team of scientists, unburdened by geographical location, generous with their time and energy, all working together to solve one colossal problem. One of the first manifestations of this synergy was the gift to CUFOS of "UFOCAT," a computer database of UFO cases provided by David Saunders.

The twofold mission of CUFOS was expressed in the letterhead: the new organization would be dedicated to *research* and *education*. "The unexplained content of many UFO (unidentified flying object) reports from credible witnesses in many parts of the world defines a problem and presents a significant mystery," read the introduction to the CUFOS mission. "Any phenomenon which has occupied the thought of so many people for so many years is surely a subject worthy of serious scientific study."

Physically, the center was nothing more than a post office box in Northfield, Illinois, and Hynek's home office in Evanston, but plans were being made, as this list of the center's five principal objectives makes clear:

1. To operate "UFO Central," a clearing house for UFO reports, with a 24-hour toll-free telephone line;
2. To pursue a rigorous study and analysis of the entire phenomenon;
3. To provide bulletins and technical reports and be a reliable source of information about the phenomenon;
4. To assist and guide correlative studies through international symposia and conferences;

5. To aid in coordinating the efforts of those in many different fields of study that can be applied to the problem . . .

The center had, within its first three months of operation, begun recruiting case investigators, adding new sighting reports to
UFOCAT, planning research projects using the data from UFO-
CAT and other sources, and establishing the beginnings of an
expansive UFO library and publishing service.

A letter Hynek sent along to Northwestern's vice president of
development in November 1974 revealed that the Center for UFO
Studies was, after only its first year, "being considered a serious
scientific effort."[21] Citing its involvement in two planned news
specials and upcoming feature articles in both the *Chicago Guide*
and the *FBI Bulletin,* Hynek made the case that CUFOS was already scoring some significant successes (in a later interview he
said that the 800 number started ringing regularly once the FBI
article had broken the dam). Furthermore, he mentioned that he
had just been recruited to be a consultant to NASA's Goddard
Space Flight Center, where he would be working with a committee that would be mapping out the course of America's space program for the next thirty years.

These accomplishments made it puzzling to Hynek that the
university did not want him to mention his position at Northwestern and his UFO work in the same breath. "I personally resent the
implication that the subject is sheer nonsense and that anyone associated with it is a crackpot (speaking bluntly!)," Hynek wrote.[22]

"I could go on with the growing interaction of the Center with
scientific bodies and laboratories around the country," he went on.
"For example, I have been asked to give a colloquium about the
Center for the University of Maryland Astronomy Department
and at the Sandia Labs in Albuquerque in the near future. Oddly
enough, I have been asked to give colloquia on the UFO subject at
several astronomy departments around the country but never here
at Northwestern! Yea, verily, the prophet is not without honor,
save in his own country!"[23]

It seemed to Hynek that Northwestern did not know what to do with its celebrity professor.

"There is no implication on the part of the administration of Northwestern University that the scholarly work in which you and your colleagues from other universities are engaged is a 'crackpot' activity," wrote Provost Raymond Mack in a contorted December 9 response to Hynek. "There is no reason for Northwestern not to receive credit on the coast-to-coast television speeches you mention for the fact that you are a member of our faculty, but there is every reason for us not to claim that the Center for UFO Studies is an organizational part of Northwestern University."[24]

The sentiments of the student body, on the other hand, were unerringly positive, in part because Hynek was now offering a class called UFOs 101, the first UFO class offered for credit at a major university (thirty-five students had enrolled and were studying how the news media covers UFO events). Despite having given up his chairmanship of the astronomy department in 1975 and having made no secret of his plans to retire in the coming years, Hynek's regular appearances in the *Daily Northwestern* campus newspaper made sure that his visibility and popularity remained high.

One article identified Hynek as "probably the leading Northwestern professor on the international talk show circuit" and boasted of his recent appearances on "the *Today* and *Tomorrow* shows, Dick Cavett's program, and in countless syndicated talk formats. He has also been televised in Australia, Canada, England, France, Italy and Japan."[25] One of his most recent appearances, the article noted, had been on a Canadian TV show hosted by the famous psychic known only as "Kreskin," in which the mentalist "apparently convinced 14 guests that they were seeing UFOs."[26]

A notice in the *Summer Northwestern* from August 1977 advertised a "Major Faculty Address" to be given by Hynek at the Tech Auditorium, saying, "Normally, you should stay away from anything of this nature. But the speaker is Dr. J. Allen Hynek, author of *The UFO Experience* and *Edge of Reality*. Whenever Hynek gives this lecture, it's before a full house. If you are at all interested, go. Hynek is great, and he won't be around much longer. (He's leaving NU.)"[27]

Kevin Leonard, university archivist at Northwestern's Deering Library, was a student of Hynek's in 1974. Hynek didn't always talk about his UFO work in class, Leonard recalled, but neither did he shy away from it, as an incident with a "streaker" proved beyond all doubt.

"Hynek had a wonderful reputation on campus; everybody knew about the guy," Leonard recalled. "I was taking his 'Highlights of Astronomy' course. We were sitting in this large auditorium because 'Highlights' was a very popular class. In the middle of Professor Hynek's speaking, we hear the doors creak open from the back of the room and a loud *flap, flap, flap. . .* He stopped speaking and he looked up, and a look of incredulity appears on his face. All the heads turn to the source of the sound, and there was a naked male, painted green from tip to toe, wearing scuba diving flippers, with a rubber mask on of what you might assume to be a space alien.

"He came hopping and flopping and stumbling down the stairs, and the room was silent apart from the sound this guy is

Dr. Hynek lecturing to a packed astronomy class at Northwestern University. Note the references on the blackboard to two hot-button topics: "Life on Other Worlds" and "Swamp Gas." *(Courtesy of Northwestern University)*

making . . . ," Leonard remembered with a grin. "The guy got all the way down to where Dr. Hynek stood at his lectern and looked at Hynek. He crooked his neck, like he was examining Hynek, walked around him, and I don't know where he had it but he had a small camera, and he took a few photographs of Professor Hynek and then bolted out the side door. As fast as you can run with flippers on, that's how fast he was going!

"The room erupted into cheers and laugher, and Hynek started to laugh . . . He just thought it was really funny," Leonard said. "After everything had subdued, he continued his lecture: back to work! It's a memory that's just burned into my brain."

When a reporter later asked Hynek why he thought the "alien" took his photograph, Hynek quipped, "To take it back to Mars, I suppose, as an example of a humanoid."[28]

More love from the student body was evident in the program of the 1974 Women's Athletic Association–Men's Union campus variety show. Skit #14, entitled "The UFO Experience," featured a cast of seven: one actor portraying Dr. J. Allen Hynek and the other six playing Martians. A photo of the production in the Northwestern archives depicts a goateed, bespectacled, but unusually tall "Hynek" standing at his telescope, confronting a group of decidedly nonthreatening Martians who look as though they just wandered over from the set of a 1950s science fiction movie by way of a Mardi Gras parade. Hynek appears taken aback, perhaps because the Martians seem to be singing to him.

These antics did nothing to endear Hynek to Sagan, whose path he continued to cross in unexpected ways. Although Hynek tried his best to maintain cordial relations, whenever the two shared a podium there was potential for friction, if not out-and-out fireworks.

A surprising letter to Sagan, written in anticipation of a November 1975 futurism conference in Chicago—featuring film screenings, "rap sessions," and panel discussions "designed to explore the question of where we have been and where we are going"—revealed that this situation was in some ways wearying to Hynek: "There is one point about our forthcoming meeting at Oasis and the possible television appearance. I would like to propose a sort of mutual non-aggression pact, because I think any confrontation (which the TV people would dearly love) would, I believe, be based on an unfortunate misunderstanding. I do not, and have never, supported the idea that UFOs were nuts-and-bolts hardware from some very distant place.

"It is not only natural for any culture to interpret a new phenomena [sic] in terms of its own present technology, but understandable," Hynek continued. "My whole point is, and always has been (as expressed in my writings, not as I have been quoted) that the UFO phenomenon is real, whatever it may be, after making due allowance for the four cases out of five which prove to be nonsense. Perhaps we disagree at this point, but certainly not on identifying UFOs as hardware from space.

"If this is clear between us, then we can have some fun discussing the subject."[29]

CHAPTER 20

HYNEK VS. SAGAN

SAGAN IGNORED HYNEK'S REQUEST. There would be no truce, no civility, no fun. There would be blood.

Only one journalist seems to have covered the debate between Hynek and Sagan at the futurism conference that took place on November 1, 1975, at the Chicago Hilton. The account of the event was published in a little-known UFO magazine that was published only quarterly; in fact, it took a full five years before the story ran. Add to that the fact that there were only about two hundred conference attendees present, and the event must surely go down in history as the most momentous event in UFOlogy that no one has ever heard of.

The drama took much of the day to unfold. Both men gave solo presentations earlier in the schedule, in which each skillfully established his point of view regarding science, research, and the unknown. In his address, Sagan took the time to comment on "borderline, marginal and pseudoscience which afflicts us today."[1]

"Some are more silly than others. All of them have this marginal character," Sagan pronounced. "I would put in this category: UFOs, the works of Erich von Daniken and other ancient astronaut enthusiasts, Bermuda triangles . . ."

He demolished the theory that aliens had visited Earth in ancient times and had influenced human development. It was an odd attack coming from a man who had hypothesized over and over again that aliens could have visited Earth a million years ago, but Sagan doubled down, declaring that ancient alien stories were demeaning to human beings because they suggested that we hapless humans couldn't make it on our own.

Hynek, in his address, asked, "Isn't it only natural for any culture to interpret new, empirical phenomena in terms of its own technology? We have modestly gone into space, ergo, 'UFOs are automatically someone else's spacecraft.'" He went on to say that he cannot settle for the "off the shelf hypothesis—'hardware from space.'"[2]

"I think we might best start," Hynek said, "with the fact that the most ardent skeptic cannot controvert. The incontrovertible fact is that reports of UFOs exist. It is also a fact that reports of UFOs continue to be made almost daily in all parts of the world by all sorts of people from all walks of life and, actually, of all degrees of education, training and culture."

After discussing what he thought UFOs weren't and what they might be, Hynek explained the importance of making not one but numerous theories to explain such things as "the unrealistic acceleration, for one; the simulation of zero mass; the isolation in space and time; a change in shape, general noiselessness; the seemingly purposeful *non*-interaction with humans; the apparent manifestation of intelligence.

"And I think we have to be open to the possibility that we cannot fashion a hypothesis yet," he concluded. "There may be things that we just don't know about yet."

That was round one. Then came what the reporter labeled, "Showdown."

When it came time for Hynek and Sagan to share the stage, "they sat with arms folded, at opposite ends of the forum like negative and positive poles."[3] Sagan led off with his expected dismissiveness, declaring that those UFO reports Hynek was crowing about could be explained away by "the full range of human misperception plus the full range of things in the sky of not terribly exotic nature" (this despite the fact that an hour or two earlier he had condemned ancient astronaut theories because they sold human beings short).

"My question is," he added with a flourish (and perhaps a flip of his hair?), "where have all the angels gone?"[4]

The argument Sagan proceeded to put forward was astounding. He declared that because Hynek's ideas were so fantastic, they demanded more rigorous research and "hard evidence," but then insisted that Hynek's ideas were not worthy of research dollars because they were so fantastic. And Sagan himself reserved the right to declare whether any "hard evidence" Hynek may present was actually, well, *hard*. To wit: "Perhaps someone will say, Isn't the Betty Hill star map an example of hard evidence? And I would say no."[5]

How simple Sagan's world was.

When Hynek responded that Hill's star map remained controversial, Sagan resorted to his trademark ad hominem attack: "A lot of the pseudo-science that I talked about before has this one advantage of it. A minimal intellectual effort is necessary to understand it."

This might have been the moment Hynek took the gloves off, if not for his own sake, then for the sake of the "minimal intellectual effort" of Betty Hill, Marjorie Fish, and the hundreds of UFO witnesses he had met over the years, but even after Sagan's swipe, Hynek behaved as though a truce were in effect.

Because of this, Sagan saw fit to continue in the same mode. He dismissed the preponderance of UFO reports with, "The fact that people talk about it doesn't mean that it really happened." He flattened Hynek by saying he thought Allen was providing an important service, but then implied that it was more a *social* than a scientific service. He reversed himself and admitted that UFO research *should* be funded, but only after the research had proven to be fruitful, a standard that would surely come as a surprise to the thousands of grant-funded scientists whose important, valuable, *worthy* research is, was, and will always remain inconclusive.

But Sagan reserved his coup de grâce for the climax of the discussion.

"I predict," Sagan proclaimed, "that if and when you ever get a really good case that involves hard evidence, there will be no lack of federal funds."

Hynek motioned to a massive study of physical trace evidence sightings compiled by CUFOS and said, "There are 800 cases in this—"

"I said **good** cases," Sagan cut in. The audience laughed.

This recalls a favorite strategy of Project Blue Book to get rid of troublesome cases wherein the investigators or program chiefs could pronounce a case unexplained, then file it as "explained" because it had been explained as unexplained.

This trick was clearly familiar to Sagan: set an arbitrary standard for what makes a "good case," then declare that there are no good cases because you alone get to interpret—and change—the standard you've established for what makes a good case.

Hynek calmly reminded Sagan yet again that he was talking about physical traces, *not* physical spacecraft. "Okay," Sagan responded, "provide me with physical evidence of something which, on its internal evidence, is extremely exotic. I mean, all I'm saying is that when there are good cases, there is money for them." Fair enough on the surface, but also fairly meaningless, as he had already made it clear that he and he alone got to say what cases were good cases, and he was predisposed to find that none of them were good.

"You don't want to throw the money after bad cases," Sagan went on disingenuously. "You want to wait for the good ones. When there are good ones, there'll be studies. No question about that."

"If somebody will listen!" Hynek said.

The reporter described the two men as though they were combatants on the battlefield, gladiators in an arena. If that's true, who was the victor?

Sagan clearly knew one of the key rules of debating: act as though you're winning, even if you're not. His aggressive tactics always seemed about to lure the more thoughtful Hynek into logical and semantic traps, but Hynek wasn't so easily manipulated. Although, as his son Paul pointed out, "he bemoaned the fact that he wasn't razor sharp with sound bites," he knew what kinds of attacks to anticipate, he knew how to most effectively make his own case, and he knew that some portion of the audience who appeared sympathetic to Sagan's argument really weren't so sure. He also likely knew that deep down inside Sagan's snide bravado there may have been a smoldering fear that someone, at some point, would realize that on many key points his beliefs were actually not altogether opposed to Hynek's. After all, the aliens Sagan believed had likely visited Earth a million years ago had to have come *from* somewhere, possibly another planet, and they had to have had *some* mechanism by which they made the trip, possibly a spaceship.

It must be further pointed out that Sagan's own pet cause, the Search for Extraterrestrial Intelligence (SETI) Institute, has been voraciously gobbling up research dollars since 1984 and has absolutely nothing to show for it. This puts it, by Sagan's own standards, on a par with Hynek's work, and thus unworthy of funding. Yet the SETI Institute's Carl Sagan Center today operates with the generous financial backing of NASA. Adding to the irony, Sagan's popular 1985 science fiction novel, *Contact,* argues strenuously that SETI research must be funded despite its ongoing failure. Surely, this suggests that Sagan held some deep, but perhaps unrecognized, admiration for Hynek and his quest.

Hynek's plaintive worry that no one was listening would come back to trouble him periodically, but it was never grounded in reality. There was, as we shall see, always someone listening.

There is no record in Hynek's papers that he ever proposed a truce with Sagan again.

CHAPTER 21

CLOSE ENCOUNTERS

FRENCH ASTRONOMER CLAUDE POHER considered himself "the opposite of a 'dreamer'"[1] and admitted that he, like many others, was a UFO skeptic at first. And yet when Hynek first met him at Cape Kennedy, Poher had just come away from reading the Condon report with a strong intuition that there was something to the phenomenon.

What changed his mind?

"The author first met this problem in 1969 during one of his numerous voyages in the States for a French experiment in Skylab," Poher said, writing about himself in the third person. "A well-known American scientist showed him a number of well documented UFO observation reports that the author, very skeptical at first, immediately began to take to pieces in search of the truth."[2]

Newly convinced by Hynek that there was a problem, Poher worked with Jacques Vallee—now a successful Silicon Valley computer pioneer—on a lengthy study entitled "Basic Patterns in UFO Investigations," which the two presented at a meeting of the American Institute of Aeronautics and Astronautics (AIAA) in 1975. In the paper, Poher and Vallee considered a number of elements of a representative sampling of UFO encounters only to arrive at the conclusion that something of scientific interest was actually going on, and that the members of AIAA should pay heed. "For me there will be no rest until the mystery is solved," Poher proclaimed.[3] Hynek had found a kindred spirit.

The UFO mystery grew a little more complex in October 1975, when *The UFO Incident* premiered on NBC. The dramatization of the 1961 Betty and Barney Hill abduction case focused on the Hills' hypnotic sessions with Dr. Simon, presenting the event as Barney and Betty remembered it and leaving it to the viewer to decide whether the abduction had occurred. Regardless of one's opinion of the Hills' story, the movie must be credited with presenting a harrowing event in a surprisingly sober, unbiased manner that leaves the reality of the incident very much an open question. An epilogue to the movie even discussed Betty Hill's star map and Marjorie Fish's work "decoding" the map without discrediting either woman, a remarkable achievement in the annals of UFO reporting.

The UFO Incident must have primed the pump of the collective subconscious, for just a few weeks after it aired, an alien abduction was making national headlines again. This time it concerned a young logger in Arizona, who was said to have vanished into a UFO after being struck down by a beam of intense light.

Hynek related the case in an address to a gathering of professional hypnotists: Travis Walton and his six coworkers had been driving home from a long day of clearing brush in an isolated area of the Apache-Sitgreaves National Forest when they spotted an unusual light through the trees. "As they rounded a curve in the road, suddenly they all saw it and they all froze," Hynek said. "They saw this brilliantly-lighted object—I should really say they *said* they saw it . . . And they froze completely; all except Travis. Travis got out . . . and ran under it."

Before Walton could return to the truck, the object struck him down with a bolt of blue light. "It came down and zapped him," Hynek said. "He told us it felt as though he put . . . a couple of fingers in electric light sockets, and it knocked him over, flat on the ground, about 10 feet away. Well, his brave companions beat it." (Huge laughter from the audience.) "Just turned tail and scrammed. But about 10–15 minutes later their consciences as you might imagine began to hurt them and they decided to come back, because, after all, it was November, in the mountainous reaches of northern Arizona, and you don't stay out there all night with just a light leather jacket.

"So they went back but there was no trace either of the object or of Travis: gone.

"They finally decided to call the police. But the police weren't going to buy the UFO story *at all*. The police suspected foul play," Hynek said.

Walton's worried workmates all insisted on taking polygraph tests to clear their names. "Five of them passed them completely; the sixth was . . . inconclusive, but not negative . . ." Hynek said. "The police began to take things pretty seriously then, and they instituted a search, but four days went by; no sign of Walton."

As clouds of suspicion gathered over Walton's workmates and his family, Walton mysteriously reappeared. Four and a half days after he had gone missing, Walton regained consciousness alongside a road, in time to see the object disappear. He found his way to a gas station, phoned his brother-in-law and asked for help. He was found by his brother and brother-in-law at midnight, collapsed in a phone booth in a town miles from where he had disappeared, hungry, unshaven, and confused. He was shocked to learn that he had been missing for nearly five days; he said he was aware of only two hours passing.

"Up to this point it's been a multiple-witness case, with all seven of the men saying essentially the same thing," Hynek said. "Now it becomes simply a one-witness case, which is unfortunate."

Walton's family decided that Travis was too weak and traumatized from his ordeal to face the police and the media, so they attempted to keep him away from the phone and under wraps. The police had been on the lookout for any suspiciously miraculous developments in the case, however, and when Travis's brother Duane spirited him away to Phoenix, "the Walton abduction case [was] in danger of collapsing into farce," according to Jerome Clark.[4]

When they were finally revealed, Walton's conscious memories of his experience were vivid: somehow, he told investigators, he found himself inside the craft. He said he encountered three large-headed entities with mesmerizing eyes, and he picked up something and threatened them with it. They backed away and then another type of humanoid entity appeared as the walls of the room became transparent, and he could see stars through the walls. "This begins to look like *Star Trek*," Hynek quipped to his audience. This new entity led Walton down a gangplank of sorts, but "Walton apparently got obstreperous again," according to Hynek. A fourth creature arrived then and produced a device that rendered Walton unconscious. He remained this way, he claimed, until he was deposited along the road almost five days later.

According to Clark's account, the list of things that went wrong in the investigation of the Walton case is a lengthy one: when Walton's brother Duane enlisted help from a UFO group named Ground Saucer Watch, the group's leader took Walton to an unlicensed "hypnotherapist" on whom Walton ultimately walked out; in response, Clark wrote, the UFO group all but declared war on the Waltons. Physicians were given a sample of what was described as Walton's first urine after his return, but the sample was handed over by Walton's brother, so its true source could not be verified. Clark went on to report that when Jim and Coral Lorenzen of APRO came in to investigate the case, they accepted an offer from the *National Enquirer* to cover some of their expenses, thus surrendering any claim of conducting an independent, unbiased investigation, and when the results of Walton's polygraph test proved troublesome, the Lorenzens and the *Enquirer* agreed not to release the results to the public, a move succinctly described by Clark as "another bad idea."[5]

Then there was the hypnosis. In Hynek's 1979 speech to the hypnotists' conference, he explained that "regressive hypnosis did *not* work very well with [Walton]. Most of the things he told us, and told the investigators, we're not sure how much of it was from conscious memory . . . we seem to think that the original abduction part and seeing the creatures was a *conscious* memory, rather than brought out under regressive hypnosis, but *very little* other than that was brought out under regressive hypnosis."

Despite the skepticism surrounding Walton's tale, it was turned into a popular book, *Fire in the Sky,* which was turned into a movie of the same name. Even today, Travis Walton remains a popular speaker at UFO events, and these days one or two of his "brave companions" occasionally share the stage with him.

BY THE MID-1970S, Hynek had fully grown into his role as the world's preeminent UFO expert and felt freer than ever to opine with authority on any and all matters related to the UFO phenomenon. Nowhere was this more evident than in his second book, 1975's *The Edge of Reality,* coauthored with Jacques Vallee.

The book was subtitled *A Progress Report on Unidentified Flying Objects,* but it was much more than that. It was a progress report on human thinking. "The UFO phenomenon calls upon us to extend our imaginations as we never have before, to think things we have never dared think before," the authors stated in the very first sentence of the book—"in short, to approach boldly the edge of our accepted reality and, by mentally battering at these forbidding boundaries, perhaps open up entirely new vistas."[6]

This was not a book about what UFOs are. This was a book about what UFOs mean, both to the human race in general and to the witnesses in particular. Presented both in prose and in transcripts of free-flowing conversations between the two scientists, with various friends serving as moderators, Hynek and Vallee discussed everything from Contactees to Pascagoula to the "Men in Black," menacing figures in black suits who are reported to intimidate UFO witnesses. Even the poor little alien buried in the cemetery in Aurora, Texas, was given its due: "Such were the hoaxes in the good old days," wrote the authors,[7] clearly not taken in by Truthful Scully's tale of a crashed airship and a mangled Martian pilot.

In one noteworthy passage, Hynek recounted his role in the oddball 1961 Eagle River, Wisconsin, case, in which a man named Joe Simonton saw a strange craft land in his backyard and went out to investigate. Simonton was met by two men who emerged from the craft and whom Simonton thought looked "Italian." One of the men held out a container that was empty, and Simonton kindly filled the intergalactic Thermos bottle from his well and returned it to his visitors. They let him take a peek inside their craft, where he saw a third man in a galley making pancakes. They gave Joe a stack of the pancakes, then reentered their craft and flew away, leaving a very befuddled man with a very strange breakfast.

The local judge, a friend of Simonton's, made a lot of noise about the incident, and in response Blue Book sent Hynek and a team of investigators to northern Wisconsin. There, they interviewed Simonton and retrieved portions of the strangely perforated flapjacks for testing. "Those were examined and were found to be ordinary grain pancakes," Hynek said.[8] Which is not to say that he completely disbelieved the story; there were details of Simonton's report that gave Hynek pause, he admitted, but the absence of other witnesses was a severe stumbling block. "As they used to say in Roman law, 'One witness is no witness,'" he concluded.[9]

The simple fact that this case was discussed with some seriousness, much less discussed at all, says a great deal about the authors and their book. To Hynek and Vallee, any and every unexpected glimpse into a possible alternate reality or unknown realm of nature was worthy of examination, even if it suggested that the UFO phenomenon possessed a distinct psychic element in addition to its physical aspects.

A familiar name entered the authors' conversation about interstellar distances: "Carl Sagan has pointed out," Hynek said, "that 'the average distance between the stars in our galaxy is a few light years. Light, faster than which nothing physical can travel, takes years to traverse the distances between the nearest stars. Space vehicles take that long at the very least.'

"Well, how do we know?" Hynek retorted. "I mean how do we know how fast thought travels? The solution may lie in the parapsychological realm; the means of getting information I mean."[10]

In what may be the most fascinating element of the book, Hynek talked at length about how teaching astronomy at Northwestern had given him new insights into the UFO enigma:

There is a certain disenchantment with science in general among people. This doesn't extend to astronomy . . . In so many new ways astronomy continues to capture and expand people's imaginations, particularly the imaginations of young people . . .

I know that particularly the kids are intrigued by two series of things these days: From the questions we get at the observatory and in the high schools I talk to, and in freshman classes, the main question areas are the black holes, quasars, and pulsars, and UFOs, bunched together. That is what grabs them. And so, to that extent, both astronomy and UFOs are expanding imagination and consciousness.[11]

A tangible sign of this expansion occurred in 1976, when NASA bowed to the wishes of thousands of fans of a long-defunct television show and named its first space shuttle *Enterprise,* after the fictional starship in *Star Trek.* NASA's plans were to christen this shuttle, the very first of the fleet, *Columbia,* but when President Gerald Ford was inundated by letters from "Trekkies" requesting that America's first reusable space vehicle be named after their own symbol of hope and optimism, Ford (perhaps recalling the intensity of public opinion surrounding the swamp gas incident) wisely directed the agency to give the ship this suddenly popular and relevant name.

Despite the fact that *Enterprise* was a test vehicle and never reached Earth orbit, the naming was of profound significance to science and science fiction fans of all stripes, but most of all to Trekkies, fans of the show that had disappeared from the airwaves some seven years earlier but had only grown in popularity in reruns. It would perhaps be redundant to point out that the naming was equally important to NASA scientists and engineers, because, in truth, so many of them actually *were* Trekkies and had chosen their careers because of *Star Trek.*

HYNEK'S FIRST BOOK, meanwhile, began working its way into the collective subconscious when it was adopted as a "how-to" manual for what would become one of the most successful motion pictures in history. What would ultimately become one of Hynek's greatest successes started out as a series of painfully awkward exchanges between Evanston and Hollywood.

> *Dear Mr. Spielberg,*
> *I am very interested in your forthcoming production,* Close Encounters of the Third Kind, *because apparently the title has been taken from my book,* THE UFO EXPERIENCE.[12]

So began a January 8, 1976, letter from J. Allen Hynek to Steven Spielberg, c/o Columbia Pictures, in which Hynek displayed a unique gift for understatesmanship.

> *I do not think we could attribute this to coincidence. Although I am pleased that this recognition is being given to my terminology, I would really have liked to have been informed of this rather than read about it in a national magazine!*[13]

After mentioning that he would be in Los Angeles in a few weeks' time, Hynek inquired as to whether he might be able to meet with Spielberg when he was on the West Coast.

> *Is there any chance that I could learn a little more about your production while I am in town?*
> *I look forward to hearing from you,*
> *Sincerely,*
> *J. Allen Hynek*[14]

In an awkward reply twelve days later, Spielberg apologized to Hynek for the use of his title, explaining that a friend had suggested it after reading Hynek's book. The director went on to say that he was making *The UFO Experience* required reading for everyone on his creative team.

As for the title of the film, there was no cause for concern: the producers, he wrote, had decided to call it *Watch the Skies* instead.

Astute readers will recall that "Watch the skies!" was the signature line of dialogue from the 1951 film *The Thing from Another World*. But that iconic title did not last long either: two months after Spielberg first wrote to Hynek, he had changed course and decided that he might just name the movie *Close Encounters of the Third Kind* after all. What's more, he—or, rather, a legal entity referred to as "Productions"—was ready to pay Hynek for the use of his term, to the tune of $1,000, with an additional $1,000 thrown in for the rights to use the stories in the book.

"I never thought an astronomy professor would fall so low," Hynek was later to joke,[15] but of course at that time neither he nor Productions had any idea how well the film might do.

To help ease the pain, Productions committed to hire Hynek as technical advisor for the movie at a rate of $500 per day, for a total of three days.

It's easy now to mock Productions' astonishing frugality, and to roll one's eyes at Hynek's lamentable negotiating skills (Hynek himself later quipped that if Spielberg ever made a sequel, he hoped he would be a better businessman the next time around), but in truth, both men got what they wanted out of the deal. Turns out, it may be far, far better to be associated for all time with an enduring cultural touchstone than to be merely well paid.

Inspired by Spielberg's youthful fascination with space and UFOs, *Close Encounters of the Third Kind* seemed an inevitable film for the director to take on, and he readily admitted that the idea for the movie was in his head long before he directed his first hit, *Jaws*. The film unreeled like a typical UFO film, going for thrills and scares, beginning with a seemingly unrelated string of unnerving encounters between random characters—FAA control tower operators, an artist and her young child, policemen, a somewhat dull family man named Neary who works as an electric company lineman—and strange things in the sky. Early on, it's easy to see how much inspiration Spielberg drew from *The UFO Experience;* the control tower scene is typical of many Blue Book "radar" cases Hynek described in his book, and the sequence in which police cars chase a string of UFOs at high speed through the countryside and across a state line is a direct copy of the tragic Dale Spaur case. There's even a shady government presence hovering on the sidelines to boot.

Hynek wasn't the only dreamer inspiring Spielberg, however. Neary's wife in the film calls him Jiminy Cricket, a reference to the talking and singing cricket in Walt Disney's *Pinocchio* who assured the lovable puppet that "when you wish upon a star, your dreams come true." That very song is heard more than once in the movie, suggesting that wishing, dreaming, and believing are the strongest powers on this or any world.

Accordingly, the film subtly shifts as wishes begin to come true. The government's interest in the phenomenon is—*gasp!*—scientific. In fact, not only have the scientists in the film begun to figure out the UFO phenomenon, but one of them is giving the orders to the military men. Neary, the artist, and several other characters are drawn together by a subconscious, perhaps hypnotic imperative to seek out a mysterious mountain, and the alien entities, portrayed initially as potentially malevolent, are revealed to have a higher purpose. In what may be the masterstroke of the movie, the characters and viewers realize that *the aliens seem frightening only because we can't understand their message!* As one of the humbled scientists says when the alien mothership starts teaching the earthlings its musical language, "It's the first day of school, fellas."

When contact is finally made, the aliens—and their craft—are grinning happily at us dumbfounded humans. "And then we understand why Neary has been compelled to be here, why he has striven so hard," wrote science fiction author David Gerrold in his 1978 review of the film. "It is not just curiosity, it is a deeper yearning to be complete, to know, to be a part of—all Close Encounters.

"We know that we, the audience, are complete when we see the alien clearly, when it turns . . . and flashes us the hand signal representing the five-note theme. What it is saying is 'Hi' or 'I love you too' or anything in between that the viewer may be feeling."[16]

To fully appreciate the significance of Spielberg's film, one can look at the financial returns (the film has grossed over $300 million worldwide since its release and was, at the time, Columbia Pictures's most successful film ever) or the accolades (*Close Encounters* was nominated for eight Academy Awards and came home with two: Vilmos Zsigmond won an Oscar for Best Cinematography and Frank Warner received a Special Achievement Award for Sound Effects Editing; John Williams went on to win a Grammy Award for the film's iconic sound track), or one can read Hynek's fan mail from late 1977 and early 1978, immediately after the movie's release.

Case in point, from February 9, 1978:

> *Dear Allen,*
>
> *I think it's just great that the current UFO thing, prompted by two super movies, has put you on stage front and center. I remember several years ago when you told me that the subject was so touchy that you and your fellow astronomers would meet in virtually secret session.*
>
> *May your dollars multiply as fast as the newspaper clippings!*
> *Sincerely,*
> *John Fields*

That was the same John Fields, VP of development for Northwestern, with whom Hynek had tussled over tying CUFOS to the university, but there was no rancor here. With the release of Spielberg's movie, now commonly referred to as *CE3K,* Hynek was once again "on stage," and that was almost as good for Northwestern's image as it was for the image of UFOs (never mind that Fields seemed to think the other science fiction hit of the time, *Star Wars,* had something to do with anything).

Or, as Hynek's son Paul put it: "*Close Encounters* was fantastic. That's the thing I remember that catapulted him to a new level of public attention.

"He was very pleased with the movie," Paul said. "He loved Spielberg."

The *Daily Northwestern,* meanwhile, suggested that in the wake of *Close Encounters of the Third Kind,* Hynek qualified as a world-changer, occupying the same rarified air as Albert Einstein, Charles Darwin, and Karl Marx. "While there was a time, not long ago, when UFO reporters were easy candidates for the mental ward," the paper claimed, "Hynek said . . . a recent Gallup poll indicates 57 percent of the American public and 70 percent of American astronomers believe a 'UFO phenomenon' exists."[17] If that's not a sign of Hynek's influence, it's hard to say what is.

Describing its beloved professor as an "intellectual superstar," the paper reported that Hynek had, in the previous two weeks, been to Italy, Alabama, and Nevada, all on UFO business. In the coming weeks, he was scheduled to be in Japan and Venezuela. But as busy as he was, Hynek never let his celebrity overshadow his "sincere dedication"[18] to his students. The article recounted how an interview that ran over time on ABC's *Good Morning America* made him only seven minutes late to his astronomy lecture, while on another day he could be found meeting with his students on a Friday afternoon after a long flight home from Alabama. Even more impressive, Hynek passed up an invitation to Buckingham Palace because it would have interfered with his teaching schedule.

HYNEK DID SPEND THREE DAYS ON SET, but not in Hollywood. The *CE3K* production was based in Mobile, Alabama, where Spielberg had had an airplane hangar converted into a soundstage. After the unprecedented box office of *Jaws,* Spielberg was given virtually everything he wanted for his follow-up production, so when there were no soundstages in Los Angeles big enough to house the set for the "Dark Side of the Moon" UFO landing base, he simply went to where he could produce his own soundstage that matched the sense of scale he brought to the production.

While on set, Hynek got to talking with Spielberg about making a "Hitchcock-type" cameo in the movie, and Spielberg loved the idea so much that he filmed a whole sequence with Hynek interacting with the childlike aliens who have emerged from the "mothership."

"At one time Steve had the idea that he would have all the humanoids come around and sort of get cozy, and while all the other scientists run away, I stay, and I just drop to my knees," Hynek said in an interview. "Spielberg zooms the camera to within a few inches of my face—he's always catching pictures when you're doing something strange—so I'm at the eye level of the humanoid, and then that humanoid and several others gather around me, and one of them pulls at my beard, and takes my pipe and looks at it, and sticks it up his nose, and things like that, which would have been corny as hell so I'm glad it got left out. But it was fun doing it, being on the set. Even had my own dressing room! But I'm really very glad that that part got left out.

"And now, it's just a very innocent appearance I make," he said,[19] harkening back to his portrayal of himself as the "innocent bystander" who got sucked into Project Sign in 1948.

In the shot, which lasts just over six seconds, Hynek is shown purposefully stepping forward toward the colossal mothership as all the other scientists hold back. As he approaches the intense light of the alien craft, he thoughtfully strokes his goatee and fingers his pipe before putting it in his mouth. He does not seem "innocent" at all; he seems for all the world as though he belongs, as if he were meant to be there. If one could have a film career that lasted six seconds, one could scarcely hope for a more memorable screen moment than this.

When asked if he thought the movie would change the way people viewed the UFO phenomenon, Hynek said, "I certainly hope so. I think that UFOs will stop being a dirty word . . . What I'm hoping for is that the sluice gates of the reservoir of unreported experiences will be opened."[20] The reporter observed that, despite its grandeur, *CE3K* portrayed a rather pedestrian "earthlings meet outer space aliens" scenario and astutely asked Hynek if there weren't other ways in which to present mankind's interaction with UFOs.

"I don't like it either, yet in the Hollywood movie, how would someone show hyperspace or parallel reality? How the hell would you portray that?" Hynek responded, perhaps showing a bit of temporal provincialism himself where visual effects technology is concerned. "It's just more to the point if it's just from out there, you see. [Steve] doesn't specify from where out there. There are no allusions to the idea that they're from another solar system even. They're just from out there someplace."[21]

In truth, the actual origin of the beatific aliens did not matter to the story line of the movie at all. If viewers wanted to believe that the creatures came from another planet, or from a parallel dimension, or even from Heaven, they were free to do so, and it's that element of nondenominational wish fulfillment that is one of the film's greatest strengths. "The whole idea of optimism and smiling and an encounter with an extraterrestrial intelligence being a pleasurable experience for mankind is something that Steven is really good at and most other directors are not," explained *CE3K*'s director of special photographic effects, Douglas Trumbull. "Most directors would have gone for drama/action/fear/suspense. And I really liked working on *Close Encounters* because it *was* about awesomeness, overwhelming beauty, and not monsters."

Two intriguing bits of trivia emerged from the awesomeness of *CE3K*. First was that Hynek became a question in the board game Trivial Pursuit: Silver Screen Edition: "What film saw UFO expert Dr. J. Allen Hyneck [*sic*] make a cameo appearance as himself?" The second involved a bit of international UFO intrigue about which many UFOlogists had long wondered. When the reporter for a science fiction fanzine asked Hynek whether Lacombe, the French UFOlogist character portrayed in the film by François Truffaut, was based on a real confederate of his, Hynek replied, "Right. Claude Poher. And yes, he does symbolically represent the scientific interest in the thing, so in that sense I'm a part of it."

Hynek's prediction that the movie would open up the sluice gates of unreported UFO sightings came true very quickly. In January 1977, eleven months before *CE3K* premiered, CUFOS had one full-time employee, Allan Hendry, who was fielding three or four reports a day. Fifty to a hundred cases a year went unexplained. That changed when contact information for CUFOS was printed on the movie posters and in Spielberg's book about the movie. As soon as *CE3K* opened the center was inundated with sighting reports, as well as requests for information, paid subscriptions to CUFOS's *International UFO Reporter* magazine, and monetary donations—one from Spielberg himself. According to Hynek, the movie "makes talking about UFOs more socially acceptable,"[22] and the numbers proved him right. In January 1978, a month after the movie was released, CUFOS was receiving "hundreds and hundreds"[23] of calls and letters every day.

Everyone who walked out of a theater after watching *Close Encounters of the Third Kind,* it seemed, had a UFO story to tell.

WHEN HYNEK RELEASED HIS THIRD BOOK IN 1977, he finally took Jennie Zeidman's advice and made sure the two biggest words on the cover were "Hynek" and "UFO."

The Hynek UFO Report was both a reach backward and a reach ahead. The book examined the vast trove of data from the twelve thousand or so case reports from Project Blue Book, dissected and critiqued the way Blue Book bungled the investigations of those reports, and then applied Hynek's own findings to his more modern sensibilities in the hopes of reaching a new level of understanding of the phenomenon.

What emerged was an almost comic encyclopedia of missed opportunities, tragic misjudgments, blatant subterfuge, and refusal to face facts. The book didn't so much reveal the existence of a government cover-up as of a government foul-up, brought about by rampant intransigence and ineptitude by the air force, whose UFO mantra, "It can't be, therefore it isn't," dictated that all worthwhile evidence be dismissed and discredited.

In a fascinating illustration of the way reporting bias can look like a cover-up, Hynek explained that CUFOS rarely received reports of "high strangeness" cases from police departments, because the police thought the cases were too unbelievable. He said that when he would query the police after the fact, the usual response was, "Oh, we wouldn't bother you with stuff like that!"[24]

"It seems likely," he wrote, "that a similar 'screening' process took place at Project Blue Book."[25]

In the book's epilogue, Hynek made a powerful plea: "The stage is set for this new adventure into uncharted fields; there exists today a growing number of scientifically and technically trained persons who are ready to devote their time and attention to the whole matter of the nature of UFOs, and to follow wherever the search may lead."

COMPARED WITH THE AVERAGE HUMAN BEING, Hynek had spent a considerable amount of his life looking up into the sky. It only follows, then, that he had seen more UFOs than the average human being: he had seen two, in fact—one from the window of a commercial airliner and another in the sky over the family cabin in Ontario. "I have seen, on two occasions, two things which satisfy the definition of UFO," he told a radio interviewer in 1981. "It was obviously an object, it was flying, and has remained unidentified as to this day. But I've never had a close encounter, that is never anything *really* close-by that I could say, 'My gosh, this is really something!'

". . . I feel somewhat left out."[26]

CHAPTER 22

ARIZONA

WHEN HYNEK GAVE HIS FINAL LECTURE AT NORTHWESTERN IN 1978, ending a teaching career that had spanned forty-three years, the unbounded affection of his students was made abundantly obvious. As Hynek expounded on the probability of intelligent life existing elsewhere in the universe, a banner was unfurled from the upper balcony of the crowded auditorium. Amid pictures of rockets, ringed planets, and a smiling space creature, the banner read, OUR LAST CLOSE ENCOUNTER WITH J. ALLEN BABY.[1]

The affection went both ways, as Hynek confessed to a reporter for the campus newspaper: "I've always felt I should have paid the university to be here rather than the university pay me," he said. "It's safe to say that now."[2]

There may have been just a trace of guilt behind Hynek's admission; after all, he was metaphorically leaving his wife for his mistress, retiring from Northwestern to devote himself full-time to CUFOS. In this he was inspired by Claude Poher, who had revealed to Hynek that the French space agency was forming an official UFO study project.

"This group is existing from the first of May 1977 and I am placed at its head," Poher wrote to Hynek. The Groupe d'Étude des Phénomènes Aérospatiaux Non-Identifiés (GEPAN) would consist of a dozen scientists from different fields and government agencies, with an equal complement of CNES engineers, Poher explained. "I hope it will change many things," he wrote.

"I would like to see a very close cooperation between CUFOS and G.E.P.A.N., with no Secrets or mysteries," Poher continued. Hynek could hardly let such an opportunity go to waste, especially since NASA had recently turned down a request by President Jimmy Carter to form its own scientific UFO research program. "We're anxious to cooperate with GEPAN," he wrote back to Poher. "We're with you the whole way!"

"The implications . . . are staggering for UFOlogy," Hynek wrote in the *International UFO Reporter* (*IUR*), the official CUFOS journal. "Arrangements for close cooperation between G.E.P.A.N. and the Center for UFO Studies are already underway."[3]

A later issue of *IUR* reported that GEPAN had completed a study of eleven UFO cases "of high credibility and high strangeness" and found only one of them explainable by "conventional" causes. "The investigations were textbook models of how such investigations should be carried out," *IUR* reported. "In 10 of the 11 cases, the conclusion was that the witnesses had witnessed a material phenomenon that could not be explained as a natural phenomenon or a human device."[4]

The same article also announced the surprising news that Poher had stepped down as director of GEPAN. "He has done everything he could with the methods at hand" was the only explanation given.[5] Sadly, GEPAN soon experienced its own financial woes. Falling victim to political pressure to stop wasting taxpayers' money, the French government stripped it of funding and resources, and in the end CUFOS survived GEPAN.

HYNEK RODE THE WAVE of *Close Encounters of the Third Kind* far longer than anyone might have expected. By 1980 CUFOS had grown in manpower and moved out of the Hynek house and into office space a few miles away. A new twenty-four-hour hotline for the general public complemented the toll-free line for police departments and public safety agencies. CUFOS even counted comedian and actor Jackie Gleason as a paid subscriber to the *International UFO Reporter*. The organization now had two part-time paid staffers and 150 unpaid investigators, and new chairman John Timmerman boasted that the center had, since its inception, investigated "87,000 reported UFOs and unexplained phenomena throughout the world."[6]

One case in particular captured Hynek's imagination and caused the greatest stir within CUFOS since the Coyne case. On August 27, 1979, at 1:40 A.M., Sheriff's Deputy Val Johnson, of Marshall County, Minnesota, was out on patrol in his brown Ford LTD squad car. Johnson, age thirty-five, was driving along a flat, featureless area devoid of buildings and traffic when he spied a bright light over two miles ahead, in the midst of a stand of trees.

The light was too high to be from a car, but it could have been on a crashed plane hung up in the trees. Johnson decided to investigate and approached the light at 65 miles an hour, noting that it did not seem to illuminate the surrounding area. As he drew within a mile and a half of the trees, the light accelerated directly at him—an instant later, Johnson said, the light engulfed the car and was actually *in* the car with him. "The Deputy Sheriff was knocked unconscious for some 40 minutes," Hynek reported, "and the last thing he heard was the windshield shattering and the brakes locking."

Johnson came to and realized that his car had stalled after skidding across the northbound lane of the highway. His eyes hurt and his vision was affected, so he radioed in for assistance. "I opened my one eye . . . and noticed that the red rectangular engine light was on, denoting that the engine had quit but the ignition key was still in the operating position," Johnson recalled in an interview.[7] The responding deputy found Johnson shivering and suffering from shock. He had a red bump on his forehead, apparently from striking the steering wheel or the windshield (he had not been wearing his seat belt). The doctor at the hospital diagnosed Johnson's eye problem as "mild welding burns"; he was treated and sent home.

When Johnson's car was recovered it was found to be damaged

in unusual ways: one headlight on the driver's side was smashed; there was a small round dent in the hood, also on the driver's side; the windshield was cracked in front of the driver's seat; two antennae on the car were bent over; and the red roof light on the driver's side had a hole in it and was partially knocked off its mount. The dash clock had lost fourteen minutes and, strangely, so had Johnson's wristwatch. The next day, an ophthalmologist found that Johnson's eyes, while having improved dramatically overnight, showed signs of having been exposed to ultraviolet radiation.

This was a case—a rare one in this era—where everything was done right. Sheriff Dennis Brekke called the CUFOS hotline immediately, and CUFOS chief investigator Allan Hendry was sent to Minnesota the next day to investigate. Hendry interviewed Johnson and found him to be as no-nonsense as Lonnie Zamora had been. He was serious-minded, kept meticulous logs of his patrols, and was considered highly intelligent and trustworthy by his sheriff and coworkers. "What he's seen he's seen. None of us are trained to explain things we're not used to seeing," said a sympathetic Sheriff Brekke.[8]

Hendry determined that there was no air traffic in the area that night, and both Hendry and the deputies came up negative when they tested the car and the roadway for radioactivity. Skid marks on the highway did not suggest the presence of a second automobile, but then what caused Johnson's two-ton Ford to be rotated 90 degrees in the middle of the lane? And what caused the puzzling physical damage to the car? Hendry's investigation was so exhaustive that he arranged for an expert from the Corning Company to examine the lenses of Johnson's glasses, an expert from Ford Motor Company to examine the car's windshield, and engineers at Honeywell to examine the bent antennae and broken lamps. The Ford engineer found that there were four individual cracks in the windshield, one of which had been caused from the inside, and three of which were caused from the outside. All four cracks were caused by mechanical forces, not heat, and all were incurred within a span of a few milliseconds. The Honeywell engineers could find no apparent explanation for the two bent antennae, but found that the broken headlight, roof light, and dented hood were likely caused by impacts from projectiles, possibly loose gravel from the shoulder. Corning's findings were inconclusive.

Hendry had to allow that most of the damage to the car could have been caused by perfectly ordinary causes—that is, flying road debris and Johnson's head striking the windshield (although it did not follow that the damage necessarily *was* caused by these things). None of this explained the bent antennae, however, or the missing fourteen minutes from the two clocks, or the fact that ultimately three physicians agreed that the burns to Johnson's retinas could not have been caused by natural light. The case was a true "unknown."

As for Johnson himself, he came out of the investigation unscathed but uneasy. "It's all hard for me to justify," he told the press at the time. "As a police officer, I work with logic, truth and facts. On this case, the facts do not necessarily represent a logical explanation. That's a bit frightening."[9] In a 2015 interview, however, Johnson was reluctant to reopen the case, telling a reporter that "I saw a ball of light . . . I drove toward it, and suddenly it was in the car with me. It's unexplainable, and will remain so. I'm happy with my mental stability."[10]

Despite Johnson's desire to leave it in the past, his case has achieved immortality of a sort: rumor has it that a dramatic UFO story line in the second season of Joel and Ethan Coen's TV crime farce *Fargo,* set in Minnesota in 1979, was inspired by Johnson's experience. Executive producer Noah Hawley tiptoed around the rumor when asked about it in an interview: "Look, the Coens' universe is very much a place where you have to accept the mystery and figure out . . . does it mean something? Does it not mean something? . . . It'll add up for most people, probably."[11]

To this day, Val Johnson's squad car rests in the Marshall County Historical Society Museum, the damage from its 1979 Close Encounter of the Second Kind left intact. It is, arguably, exactly the kind of hard evidence Carl Sagan challenged Hynek to produce.

A few years after the Johnson case, Hynek again gave a status report on CUFOS to the *Daily Northwestern,* which never seemed to lose its fascination with the university's illustrious professor emeritus. Hynek, for his part, seemed to relax and open up for student reporters in a way he rarely did for professional journalists. In a wide-ranging conversation, Hynek recognized that "UFOs are a subliminal theme in society," appearing in "many of today's video games, movies and in rock music,"[12] although he didn't seem aware of the part he himself had played in that development. "It's a part of the culture," he said, then added: "It's a new form of religion with some people—a dissatisfaction with the old-time religions in which people are looking for a scientific twist.

"I would not be prepared to defend the thesis that UFOs represent visitors from outer space," he went on. "Indeed, I think the answer may be even more interesting than that. I think the answer will be very exotic and beyond our imagination; possibly something that we would call paranormal."[13]

Then, just when it seemed Hynek might float off on a blue beam of light, the reporter brought him back down to earth in a discussion of his career in education, mentioning that he had just taken the prestigious post of astronomy editor for *Science Digest* magazine. "I never once felt in my life that I was working for a living. I do what I have fun doing," he said. "In a sense I've retired, but all I've done is shift gears."[14]

At this juncture, after all the talking, debunking, investigating, discrediting, proclaiming, and desk pounding, what did anyone really *know* about UFOs? It was undeniable that UFO reports existed and that there were certain near constants among UFO reports and UFO experiences, both geographically and over time. It was undeniable that there were strong similarities between alleged UFO occupants and mythical, supernatural beings from our past. It was undeniable that UFOs were often seen and reported by very responsible witnesses, who had little to gain and much to lose after telling of their experiences. It was undeniable that UFO reports came in periodic waves, or flaps. But, in 1980, there was a peculiarity of the UFO phenomenon that was rarely, if ever, mentioned, perhaps because of the host of unanswerable questions it raised: the phenomenon appeared, at times, to consciously frustrate and undermine human beings' attempts to understand it.

Hynek seemed to have some awareness of this when he wrote his editorial for the May/June 1982 *International UFO Reporter.* In this column, he considered the phenomenon of the UFO flap, and the fact that flaps seem to come at the worst possible times for UFOlogists. Take, for example, the 1973 flap . . .

"It was a memorable wave, containing within one week two notable 'classic' cases (the Pascagoula 'fishermen' and the Coyne helicopter cases) and a great many other Close Encounter events," he wrote. "It was largely out of frustration that so little was able to be done by ufologists to profit by such a wave, both from the standpoints of public education, public relations, and scientific study, that the Center for UFO Studies was founded."[15]

The flap ended before any serious investigative effort could be organized, and Hynek blamed the failed response on two familiar culprits: a lack of coordination between UFO investigators and a lack of funds.

"HOW DO WE HANDLE THE NEXT FLAP?" he asked. "In any event, we must do better than last time! The probabilities are high that there will be another flap and we certainly don't want it to have another frenzied, disorganized, 'chicken-with-head-off' reception."[16]

But there was no new flap in North America, not in Hynek's lifetime. It was as though the phenomenon had been warned that Hynek would be ready for it next time, so it made certain there would be no next time.

"The strange fact of the matter—and it may be the most important development on the UFO scene since the early flaps of the 1940s—is that UFOs are a lot less with us than they have been in a long time," confirmed Jerome Clark, who took over the editorship of the *International UFO Reporter* from Hynek in 1985. He noted that "most newspaper articles these days are about UFO buffs, not UFO sightings."[17]

Clark speculated on the causes of the UFO drought, theorizing that "the UFO phenomenon is changing in some way."[18] If UFOs were real, he wrote, then the intelligence behind them had changed tack; if UFOs were figments of witnesses' imaginations, then the social forces behind UFO sightings had either ended or changed in some fundamental way.

Similarly, Hynek was aware that the apparent physical nature of UFOs, or lack thereof, seemed designed to "outrage common sense." He likened it to the seemingly impossible fact that light is both a particle and a wave at the same time. Physics says that can't be, but there it is. "There's a certain dualism to the UFO phenomenon. On the one hand, the UFO seems to be material: it leaves physical traces, it can be photographed, it appears on radar, it creates electromagnetic disturbances, people perceive it as physically real," Hynek said. "And yet, it has what you might call psychic or paranormal aspects. For instance, we have cases in which the thing seems to dematerialize—to just disappear. Or conversely, to appear."[19]

Is it any wonder, then, that so little seemed to have been accomplished?

There were other reasons for the apparent ineffectiveness of UFOlogy. Around the time Hynek started devoting his full energies to CUFOS, strange news broke among UFO circles about a major event in New Mexico that would come to shine a brilliant spotlight on the inadequacies of modern UFOlogy. It didn't get much attention from the media because it actually dated back to July 1947, less than a month after Kenneth Arnold spotted his flying saucers and kicked off the "modern era" of UFOs.

The story did, however, get a lot of attention from UFOlogists, despite the fact that what came to be known as the "Roswell Incident" hardly counted as an incident at all.

In early July 1947 (the exact date has been guessed at and revised multiple times), a rancher named Mac Brazel found what seemed to be wreckage of some flimsy metallic object spread out in a remote area of a ranch outside Roswell, or so the story goes. Brazel reported his find to the police when he got back home, and the police reported it to Roswell Army Air Field. In response, an intelligence officer named Jesse Marcel retrieved some of the debris from the ranch and brought it back to the base. Despite the fact that Brazel did not see any flying object or witness any crash, and despite the fact that there was nothing about the debris that suggested it came from a saucer-shaped aerial craft, the airfield press officer issued an official statement that the army had got hold of a flying saucer.

This sensationalism did not sit well with the top brass, and in a very short while the embarrassed officers at Roswell were forced to admit that the announcement had been in error, and that the debris was from a crashed military balloon whose purpose was to detect disturbances in the upper atmosphere indicating that the Soviets were testing atomic weapons. Before the retraction could go out, however, the crashed saucer story had gotten just enough oxygen to give it new life.

"This story has come up so many times," Hynek declaimed in 1979, when a TV interviewer brought up the topic of saucer crashes.[20] Because CUFOS had collected "about 30 anecdotal accounts" of such reports, Hynek could not in good conscience completely discount them, but there was, he said, an essential problem with the saucer crash/alien corpses narrative. "People will come to us and say, 'My uncle was the surgeon who performed the autopsy on the things,' or 'My boyfriend or ex-husband was the pilot who brought the cadavers to Wright Field,'" Hynek explained. "But then when I ask them, I say, 'Would you be willing to sign a statement to that effect?' and they say oh, no, no, no, of course they can't; they are under security regulations." Hynek said that if he were ever to meet President Carter, he would request presidential immunity for those thirty people so they could talk. "Until these people who tell us these things are willing to stand up and be counted and sign an affidavit, to me this is still just a story."[21]

"To be honest, I don't like to talk about crashed saucers because I am in a position to mobilize public belief," he said in a later interview. "If I came out and held a press conference to say that a saucer has landed and the creatures were in deep freeze at Wright Field, quite a few people would believe me. But it wouldn't necessarily be true, and it certainly wouldn't be science . . . I won't jeopardize my reputation for the sake of a story."[22]

IN THE SUMMER OF 1984, in what must have seemed a very rash move to friends and colleagues, Allen and Mimi Hynek moved south.

"Chicago . . . is a hotbed of inertia," Hynek said to a *Chicago Tribune* reporter in August 1984, to explain his and Mimi's move to Paradise Valley, an affluent suburb of Phoenix, Arizona.[23] Hard evidence in UFO cases, Hynek explained, requires hard currency, and there was never enough of the latter. CUFOS was now scraping by on an annual budget of $40,000, making detailed field investigations like the one Allan Hendry had made in Minnesota a rarity, and Hynek felt that money might flow more freely in a sunnier climate.

"We could manage only to keep the pot simmering on the back burner," he told the reporter. "Now, in Arizona, we hope to do much more."[24]

The Hyneks were living in, and operating CUFOS out of, a "spectacular hacienda in the sun"[25] provided by a wealthy British benefactor named Geoffrey Kaye. The generous sponsor, whom Hynek had met through gold-mining entrepreneurs Tina Choate and Brian Myers, was ready to fund "a UFO research center without rival in the world,"[26] and Hynek couldn't resist the lure. He didn't intend to operate out of someone else's house for long, however. In time, he and Mimi would move to a more modest dwelling, and the new International Center for UFO Research (ICUFOR) would be housed in a more functional office environment. "We're going to rent an office in downtown Phoenix," he said. "We want a setting that's more egalitarian, more appropriate to a lab."[27]

The irony of CUFOS's financial straits was that Hynek, had he wanted, could have cashed in on his status as a UFO guru and become independently wealthy any time he chose. But exploiting his celebrity for monetary gain was not in his DNA. "He could have started a cult. He could have gotten all kinds of money," said son Paul. "He was not a money-grubber, that's for damn sure. Because he just didn't care; he was frugal because of his formative years during the Depression, but he just did not care about money."

But now there was going to be money to burn, to the tune of an anticipated $2 million in annual operating funds. Because the editorial and investigative functions of CUFOS remained in Chicago, along with the UFOCAT computer files, and the business office would remain in CUFOS chairman and treasurer John Timmerman's home in Lima, Ohio, where it had been for some time, Hynek would be free to branch out in Arizona. He would continue to serve as the largely symbolic editor in chief of *IUR,* but he would concentrate on two pet projects at the new center: the first would be to further develop UNICAT, a computer program that would identify correlations of features of UFO reports; the second would be to assemble multidisciplinary teams of scientists to "live and breathe carefully selected cases, even if it took years."[28] This effort, it was hoped, would produce a library of technical reports acceptable to the National Academy of Sciences and finally establish UFOlogy as a legitimate branch of scientific research.

"Originally we were just going to be a branch of CUFOS, operating here and sharing resources," Brian Myers recalled. "It was meant to raise up their field of operations and what they were able to do and their exposure . . . It was all positive."

The new center in Arizona would also supplant the amateur UFO groups that had made such a poor show of things with the Walton case. "I do not mean to sound unkindly, but the UFO movement today is filled basically with amateurs," Hynek said. "Most of the investigators are not professionals, and they are technically ill-equipped and lack funds. Many also are beset by preconceived notions of what UFOs ought or ought not to be."[29]

In addition, Hynek had long been concerned by what he described as "useless bickering and unproductive competition"[30] among amateur UFO organizations, which tended to hoard their case files, withholding valuable data from one another for petty territorial reasons. This was, in fact, the very reason Hynek had cocreated the UNICAT system with physicist Dr. Willy Smith, "to stop this erosion of our ufological resources."[31]

"Thus, as 1985 begins, we feel that CUFOS starts this year with a new vitality and a new promise of very interesting things to come," Hynek wrote in *IUR*.[32] Befitting the new paradigm in UFOlogy, the center even had a catchy phone number: (###) WHY-UFOS.

Some of the staffers Hynek left in charge of the Chicago operation didn't always pay heed to their founder's wishes and dictates. Where Hynek was patient and willing to play the long game, they were prone to impatience and anxious to make their mark on UFOlogy. Describing some of the power plays among the staff that had plagued the center since he left Chicago, Hynek told Jacques Vallee, "There's been some politicking when I left Cufos, of course."[33]

"Hynek had moved to Arizona, and he had no influence," recalled Michael Swords, a former CUFOS officer and retired science professor. "In the earliest days, those guys were constantly taking cheap shots at people. It was not mature behavior in that office now that the old man was down in Arizona. They were in need of adult supervision. I was brand new down there, and I just was kind of appalled by it."

According to Myers and Choate, who were instrumental in persuading the Hyneks to relocate to Arizona, there may have been a certain amount of jealousy holding sway among some of the CUFOS staffers in Chicago. These men and women had all thrown in with Hynek, but now their leader had moved far, far away and seemed to be creating a new entity to take the place of CUFOS. What's more, Myers and Choate, the new kids on the block, now had direct access to the boss, and they did not.

But Hynek had bigger problems than unruly, malcontented volunteers.

"I am completely disassociated (and I mean completely) from the Phoenix operation," he wrote in a terse announcement that appeared on the last page of the last issue of *IUR* for 1985, a mere eighteen months after he had moved to Arizona. "My connection with the International Center for UFO Research is null and void."[34]

What went wrong?

Geoffrey Kaye's largesse, for one thing, was not infinite. "Kaye doesn't really intend to fund the Center," Vallee recalled Hynek telling him a few months after the move to Arizona. "He is only providing an initial framework and startup funds," Vallee reported. "His idea is to launch a series of publications and film projects whose proceeds would support the organization, starting with a movie about Allen's life."[35]

Brian Myers was named president of ICUFOR, and he started plans to make Hynek's work accessible to a wider audience while bringing in revenue to fund the full operation of the new center. He and Choate thought big; they arranged for Hynek to supply UFO photos for a space science exhibit, somewhat blandly titled *First Contact: The Search,* at the Scottsdale Center for the Arts that inexplicably included a tribute to discredited Contactee Billy Meier, and would ultimately feature speeches by Jacques Vallee and Arizona's most famous abductee, Travis Walton. This exhibit, Choate and Myers said, was the beginning of a joint project with the city of Scottsdale to build an expansive public-private "space center" to house ICUFOR's burgeoning operations. Meanwhile, negotiations were under way for a television project based on Hynek's work, to be made in partnership with TV game show producers Jack Barry and Dan Enright, the team that brought the world *Tic-Tac-Dough* and *The Joker's Wild* and had been deeply embroiled in the infamous quiz show scandals of the 1950s. Choate also described a dramatic TV series that ICUFOR was developing with Universal Studios: "Geoffrey thought that there should be, you know, a story about Allen and the research," she explained. "Our idea was to do . . . a series based on a UFO research center headquartered in Arizona. Universal was going to come and build the UFO center and was going to leave it for us to use after the fact."

Vallee noted that when Hynek learned of these plans, he was "touched by all the attention," but that he also seemed "a bit lost and disoriented."[36]

Kaye, meanwhile, had a network of affluent friends and business partners he planned to approach for financing, and in due course he held a fund-raising event at which he introduced Hynek to his "community." "There were maybe fifteen people there that he had picked from his charities that he supported," Choate said. "That's why he fully expected all of them to write a check, and they didn't."

The funding situation became more tenuous when Hynek's health began to falter, and blood tests run in late 1984 revealed that he might have a cancerous tumor. Suddenly Kaye wasn't so sure that he was making a sound investment, and he was reluctant to continue soliciting investors. According to Vallee, Kaye made repeated appeals to him to lend his support to ICUFOR—to perhaps even step in and head the organization—but Vallee mistrusted Kaye and rejected the offers.

"Geoffrey Kaye's funding never really dropped out. It was there," explained Myers. "It was only at the end, when Allen was getting ill . . . that Geoffrey's concern was 'Well, you know, how much further do I go with this, because Allen is obviously on the way out.' It was pretty obvious his health was failing."

"ALLEN IS BEGINNING TO MEASURE the distance between his dreams and reality," Vallee wrote with some alarm after traveling with Hynek in early 1985 to lobby the French government to continue funding GEPAN.[37] Kaye met with Hynek and Vallee in Paris, and the struggle to continue funding ICUFOR was clearly an issue, according to Vallee. As if that weren't stressful enough to Hynek, the rigors of international travel were clearly taxing him as well. He seemed tired and confused much of the time they were in France, Vallee wrote, describing Hynek's rambling presentation to the French officials as all but incoherent. There was, in Vallee's eyes, a sadness in his former mentor.

The sadness only grew when Hynek's colleagues started to lay claim to his legacy. "Because Allen was in such a debilitated state, everybody was clamoring for a piece of Allen," Choate recalled. Shortly after Hynek had moved to Arizona, the officers of CUFOS had pointedly changed the name of their organization to the J. Allen Hynek Center for UFO Studies; they then sent

Choate and Myers a letter demanding that Hynek's name not be used in association with any of ICUFOR's activities. "This saddened Allen," Vallee recalled, not least because Hynek credited Choate with "bringing him to the freedom of Arizona."[38]

The true reasons for the collapse in Arizona remain murky, with Vallee's version of events often clashing with that of Myers and Choate. Meanwhile, a bread crumb trail of sorts throughout the editorial pages of CUFOS's *International UFO Reporter* in 1985 and '86 adds even more ambiguity to the sequence of events. In the January/February 1985 issue, the uncredited editorial clearly stated, "The Arizona facility, or 'CUFOS WEST,' is located . . . in Scottsdale, and is a short distance from the new Hynek residence."[39] It then went on in some detail outlining the ways the two facilities would work in tandem, while noting that "the Arizona group will observe a degree of independence." Five issues later, in the November/December edition of *IUR,* an uncredited news brief announcing Hynek's retirement from his editor in chief duties and his split with Kaye maintained that "contrary to widespread misunderstanding, ICUFOR was an entirely separate organization from CUFOS . . . It was not CUFOS' 'Phoenix Branch,' as some have alleged."[40]

Then, over a year later, CUFOS cofounder and CEO Sherman Larsen wrote the following special statement for the September/October 1986 issue of *IUR:* "We wish to emphasize that CUFOS and ICUFOR are not—and never have been—associated in any way. At one time Dr. Hynek was led to believe that through ICUFOR a well-financed program of scientifically-based UFO study would come into being. When he learned otherwise— specifically when he found that the promised money was not forthcoming, that ICUFOR personnel were contactee-oriented types with a metaphysical agenda, and that commercial exploitation, not scientific investigation, was the organizers' primary concern—he withdrew his name and support."[41]

Bill Chalker blamed the collapse of ICUFOR on what he called the "Arizona mob," recalling that "the 'dream' turned out to be a mirage. Allen had his affections for the paranormal, but probably found their take and trajectory incompatible with his own and ultimately a bridge too far, remote and adverse to his own."

"I think [Dad] got a little bit desperate and he started associating with some people who really weren't that qualified in science," said Paul Hynek, summing up the whole unfortunate episode.

"Arizona was just a bunch of bullshit."

CHAPTER 23

THE SUPERSENSIBLE REALM

BEFORE THE CRASH CAME, and Hynek withdrew from his dealings with Geoffrey Kaye, he saw an opportunity. Even with the reservations he had concerning amateur UFO groups, he recognized that each group had its strengths: upstart organization Mutual UFO Network (MUFON) had developed an unparalleled national structure; APRO possessed valuable case reports that needed to be added to UNICAT; the Fund for UFO Research (FUFOR) had, at least theoretically, funds; CUFOS had experienced researchers and investigators, as well as Hynek's decades' worth of Blue Book files; and ICUFOR had Hynek. Might it benefit UFOlogy if these disparate organizations could communicate, cooperate, and pool resources?

The idea had originated with MUFON's state director in Massachusetts, Marge Christensen, who had pitched to Hynek and Willy Smith the idea of forming an advisory group of officers from all the major organizations that would find ways for their groups to combine forces. Although he was now only an advisor to CUFOS and had no power to influence its decisions, Hynek liked the concept and wrote to Christensen that a "meeting of minds" about "how the efficiency of our efforts can be improved" would be most welcome.[1]

Despite the dearth of UFO reports in the 1980s, there were still mysterious cases here and there that captured Hynek's imagination, and pooling resources to investigate them made sense. The 1980 Cash-Landrum sighting, for example, had been studied for years but remained unsolved in 1985. The story starts out with an almost comically clichéd setup: two women and a child driving along a lonely highway at night come around a bend to see a huge object hovering directly over the road. The diamond-shaped object was emitting a brilliant flaming light that reportedly melted the asphalt on the highway and the dashboard of the car. The women stopped and got out of the car, only to watch as twenty-three military Chinook helicopters appeared in the sky, surrounded the object, and escorted it away. The women later reported suffering what appeared to be radiation burns. This Close Encounter of the Second Kind has never been satisfactorily explained, in part because of the mysterious presence of the helicopters.

The Hudson Valley (or Westchester) sightings of 1982 were intriguing enough that Hynek lent his name as coauthor to a book about the event. Thousands of people along the Hudson Valley in New York saw V-shaped groups of lights cruising silently overhead at night, sometimes hovering over bodies of water, often giving off an impression to witnesses that they were intelligently controlled and could respond to their thoughts. Although some of the sightings were explained away as small planes flying in formation, there was no getting around the fact that small planes cannot hover soundlessly. Hynek was "terribly excited"[2] by the case, in no small part because of the large number of credible witnesses reporting the same phenomenon independently.

A third puzzling case occurred in a remote rural valley near the village of Hessdalen, Norway. For years, strange nocturnal lights had been appearing over the valley, but there was never any discernible pattern to the appearances: the lights took on many different colors, they traveled at different speeds and in different directions, they appeared in different areas, and some nights they didn't appear at all. When the apparitions spiked in 1984, they became a matter of great interest to UFOlogists around the world.

Hynek himself traveled to Norway in early 1985 to take part in the investigation, but in the end his findings were inconclusive.*

WHEN CHRISTENSEN'S MUFON CHAPTER hosted a UFO forum that August in Boston, the ingredients for a new era in UFOlogy seemed to be in place. Hynek was slated to give a talk on his recent trip to France to lobby for GEPAN, while Willy Smith was there to make a presentation on how UNICAT would catalog UFO data so that it could not be ignored by the scientific community.

But when Hynek took the stage to give his presentation, he faltered. He turned to his friend David Jacobs, who was emceeing the event, and made a surprising statement: "Dave, I'm not feeling so well. I'm not sure if I can give my paper."

Jacobs asked if he could deliver Hynek's paper for him, and Hynek accepted the offer. He sat at the back of the stage as Jacobs did his best to read the address, peppered as it was with handwritten notes and corrections and French phrases. When Jacobs finished, Hynek agreed to a Q&A session, but he struggled to engage in the repartee, according to Jacobs.

"We would ask him sometimes very involved questions and he would reply 'yes.' His answers were too short," Jacobs recalled. "He answered a few more questions, but Marge had already called the ambulance. So when he gets off the stage, there are the ambulance crew waiting for him. He didn't want to go, but we told him it was for the best and we wanted to make sure everything was okay."

The next day Hynek felt well enough to rejoin his colleagues for dinner, Jacobs said, but something was not right. "He sits

---

* Recent research into the Hessdalen lights suggests that mineral deposits on either side of the Hesja River, which runs through the valley, may be interacting with sulfuric river water and acting like a gigantic battery, sending electrical charges into the air above the valley. This explanation is in some ways similar to Hynek's "swamp gas" theory of the 1966 Michigan sightings, proving that UFOlogists should never underestimate Mother Nature's capacity to surprise.

down at my table with his friend Willy Smith. Allen's okay, but he doesn't want anything to eat. Then Allen is leaving for his hotel, and the next day I hear that they found Hynek sitting in the lobby for four hours waiting for the others to arrive. It shocked everybody."

Mere weeks after the conference, Hynek was diagnosed with prostate cancer and had surgery a few weeks later to remove the malignancy. But there would be no recovery, only a slow decline. A brain scan conducted while Hynek was in the hospital revealed a brain tumor, and that, Jacobs recalled, was when "he called up everybody and told them that he was saying good-bye."

"We should have known from the signs of mental confusion," Vallee wrote,[3] but he is far too hard on himself. Cognitive and behavioral effects of a brain tumor can be very subtle and difficult to spot, even by health care professionals, and Vallee often didn't see Hynek for months at a time. It is also possible that Hynek was aware of the onset of the symptoms and did his best to conceal them from those close to him. Vallee couldn't have known, and, quite possibly, no one else could have, either. However, it's easy to see in hindsight how common symptoms of brain cancer such as confusion, memory loss, and impaired reasoning and judgment may have contributed to Hynek's uncertainty in the last chapter of his life.

Hynek's brain surgery was to be performed in San Francisco, so Vallee and his wife, Janine, opened their Bay Area home to Allen and Mimi for the duration. Vallee noted that "the old positive energy does remain in Allen's voice," but when he arrived in San Francisco on September 23, he "looked frail, walked slowly."[4]

Over a month after the operation, Hynek was discharged and he returned home to Arizona with Mimi, where he started radiation treatments. The final months of Hynek's life were spent with family and friends, resting and reminiscing and reflecting on his life and his work. Hynek's children Scott, Roxane, Joel, and Ross, living in distant locales, visited their dad and mom in Arizona as often as they could. The circumstances were heartbreaking, but the children knew how happy their father was to be living in the desert, far from the icy Illinois winters, and the visits were happy ones. Paul was fortunate enough to live with his parents in Scottsdale for the last year of his father's life, and he remembers it as an uncharacteristic yet somehow natural period of spirituality for his dad, who had been anything but religious for most of his life.

"One of the aspects of our childhood that people find remarkable is that there was not even one scintilla of religion. We were completely areligious; it was all about science," Paul recalled. But that began to change after the two cancer surgeries. "Especially when he was dying, when I was living with him and my mom," Paul said, "he became even more interested in, I guess you could say 'unusual phenomenon.' You get more interested in things when you know your time has come."

As a case in point, this short essay discovered among Hynek's CUFOS papers, with no date and no notation, reflects his abiding fascination with both mankind's place in the universe and his own:

> *Any man who says that he thinks of himself as an ordinary man is a liar. Any man that says he doesn't matter—as one individual in the galaxy—is a liar. This sounds like shades of John Donne, and probably so. There is no such thing as an ordinary man.*
>
> *There are physical and emotional characteristics common to all humans. And there are humans who are educated and creative, who travel, and who constantly ask 'why?' 'when?' 'what?,' and there are humans who live and die by the millions, in starvation, poverty and ignorance, who have never been 10 miles from home, and who don't know what a star is, and are too hungry to care. But they are not common people; they are, each one, distinctive individuals, and the paradox is that each one does matter— just as each star or planet, regardless of size or distance, does exert some influence upon the others.*

Hynek's musings on a type of spiritual gravity binding every human being to the cosmos strangely melded the teachings of his astronomical idol Johannes Kepler and his spiritual mentor Rudolf Steiner, and that may have been part of the appeal for a man like Hynek, who rarely shied away from a conceptual challenge.

In the preface to his book *An Outline of Occult Science,* Steiner examined the difficulties, by now all too familiar to Hynek, of trying to open the minds of "those who are today esteemed exact thinkers" to the idea of a "supersensible realm":

> First, no human being will, on deeper reflection, be able in the long run to shut his eyes to the fact that his most important questions as to the meaning and significance of life must remain unanswered, if there be no access to higher worlds . . . The person who will not listen to what comes from these depths of the soul will naturally reject any account of supersensible worlds. There are however people—and their number is not small—who find it impossible to remain deaf to the demands coming from the depths of the soul. They must always be knocking at the gates which, in the opinion of others, bar the way to what is "incomprehensible."[5]

Under the clear desert skies of Arizona, Hynek may well have spent many hours recapturing the feeling of being under the dome at the Yerkes Observatory, gazing at the stars night after long, freezing night, contemplating his place in the cosmos, and learning the practice of reaching out into the supersensible realm and knocking at the gates of the incomprehensible.

FOR YEARS HYNEK HAD FRETTED that the return of Halley's Comet in 1986 would herald his own death, and it was beginning to look as though he had been right all along. Jennie Zeidman, who had always scoffed at his predictions of his own doom when the comet returned, had a long-standing date with Allen and Mimi to go comet watching when Halley appeared again, so one cool night in March 1986, she found herself traveling through southern Arizona with her friends, on the lookout for the faint, wispy tail of the comet as she fought off her deep sense of mourning.

Mimi turned their Honda off the main highway onto a gravel road, where a few other cars and sky-watchers had already taken up positions. Zeidman recalled the silence of the moment and likened it to an art museum or a religious site, "where out in public there is great respect for the privacy and emotion of others."[6]

Hynek had brought along the latest issue of *Sky & Telescope* magazine and a pair of binoculars, but even so the experienced stargazer needed a few minutes to locate his quarry. "Saturn and Mars were easy," Zeidman recalled,[7] but a bright patch below Sagittarius had the threesome fooled for a few moments. As they were rechecking their star chart it appeared, just rising over the horizon, "the tail streaming away like a feather in a cap," Zeidman wrote.[8] In that moment, Hynek was reunited with an old, old friend. The spaceman born under the comet's tail seventy-six years earlier was back for one last encounter with an unusual body in the sky.

"The earth turned and the comet rose," Zeidman wrote. "And then there didn't seem to be any reason to stay, so we got back in the car and drove home.

"I could leave it there: the visit that we both knew would be the last, the despair of a good friend's goodbye—but this was not a sad trip. It completed the circle and I came away with a smile."[9]

Three more times, at least, Hynek hitched rides into the desert to get a clear view of the comet. Tina Choate and Brian Myers took him once. Paul Hynek drove his dad out to see it twice, and he remembers the second trip as being less about the beauty of the comet and more about the long shadow it cast over his father's life.

A little over a month after his last meeting with the comet, Hynek entered the supersensible realm. "Respecting my father's wishes, we did not cremate him for three days, which is a Rosicrucian belief," Paul said. Hynek did not want anything interfering with this trip.

IF THE UFO PHENOMENON is preparing us for something, then Hynek played a decisive role in that preparation.

Thirty years after his death, Hynek's friends and colleagues are still trying to understand the bewildering scope of his influence. "I wish there was another Hynek around now," said David Jacobs. "There is no person who is a major scientist with a major reputation who can say, 'We have to look at this. We have to look at the subject seriously.' Nobody can say that now. There's no figurehead, so to speak. He was the one and only."

"Allen based his pro-UFO conclusions on the kind of evidence brilliantly laid out in *The UFO Experience,* still one of the literature's seminal works," said Jerome Clark. "I doubt that any human being investigated as many reports in the field as Allen did. He spoke with a unique authority there."

Nonetheless, there was always questioning, admitted CUFOS archivist and historian Frank Reid. "He vacillated, sometimes day by day or hour by hour," Reid recalled, "[but] Hynek's vacillation on the subject was proof of his scientific integrity."

It was that integrity that inspired British UFOlogist and author Jenny Randles, to whom Hynek was both a role model and a mentor. "I do remember him saying that the thing to do was face such a possibility [of alien visitation] head-on, without presumption, and 'follow where the evidence trail leads.' This is advice I have always heeded," she said. "He reminded that you should never presume an explanation—or a lack of an explanation—ahead of conducting any investigation, though he knew that it would be tempting. But if you do, then you can easily mislead yourself and unconsciously discover only what you are expecting to find.

"He knew that most UFO sightings could be explained and you had to rigorously investigate each one and resolve them whenever you could," Randles continued. "Because this then made the unresolved cases far more interesting and more useful."

"Both of us, I think, were lured to the edge by having this attitude of openness toward the possibilities," said Hynek's old right-hand man, Bill Powers. Neither he nor Hynek liked to be told "that can't be," and as a result, he said, "you don't have to take many more steps to become a believer."

Powers also revealed that Hynek may have felt some pangs of regret concerning his career. "He started life as an astronomer interested in spectroscopy," he said. "That was his main passion, was looking at the chemical composition of stars and gas clouds in space, but he never got to do that. He became director of the observatory and got involved in UFO stuff and involved in NASA stuff and all of that, so he never really got back to his main love. Every now and then he would say something that made me realize he wished he could do that."

"Hynek was a rational person looking at an irrational subject, and he was struggling with it," said his onetime foe James Oberg. A NASA computer scientist, science writer, and NBC News space consultant, Oberg had long felt that the air force should have hired a perceptual psychologist rather than an observational astronomer to explain away UFO reports. Yet despite his belief that Hynek had been the wrong person for the job, he admitted that he and many of Hynek's antagonists felt great respect for the man. "Hynek created this attitude toward himself from others that is a tribute to him and a tribute to his character," Oberg said.

"I appreciated the person, his calm, his long silences," said Claude Poher, in what may be the kindest, most aware remembrance of all. "We both were 'astronomers' (myself astrophysicist), therefore we understood each other very well."

There remain today lingering questions and mysteries about Hynek's work. Was there a government cover-up of UFOs? Was Project Blue Book merely a PR front, concealing a much more secretive UFO study? And, if so, was Hynek a willing participant in the deception?

Hynek was aware for many years that some very intriguing UFO cases were never reported to Project Blue Book, perhaps because the reports had been intercepted and redirected to some shadowy office squirreled away in the darkest corners of the Pentagon or Langley. The fact that Captain Ruppelt had not been allowed to investigate the 1952 Washington Merry-Go-Round was a significant indicator of this, as were the air force's transparent attempts to hide its interest in the 1973 Pascagoula abduction case.

On occasion, Hynek would hear of secretly diverted investigations at meetings or cocktail parties, and one that stuck out in his memory was related to him by a Milwaukee attorney who had at one time been stationed at Ladd Air Force Base, an antiaircraft installation in Alaska. "He told me at that time he was a lieutenant, and in broad daylight, the radar picked up something, their guns locked onto it," Hynek reported. "He didn't see it, but ten of his men saw this thing and they were waiting on orders from him to shoot. By the time he came down the thing had zipped away, *but,* he had the good enough sense to get all ten of those men to sit down and write their descriptions of what they had seen. He then compiled them independently, put them in a nice package, sent them on to the Colonel, thinking that it would go on to the Pentagon. He *never* heard a thing about it, and he asked me to see whether any record of it was in Blue Book, and it wasn't."[10]

It's a compelling story, but are missing reports an indication of a cover-up, as many still claim?

"Well, there are two kinds of cover-ups," Hynek said. "You can cover up knowledge and you can cover up ignorance. I think there was much more of the latter than of the former."[11]

He had long taken for granted that because the U.S. government is so intelligence-happy, there must surely be someone somewhere keeping an eye on the UFO phenomenon. That could certainly explain the missing report from the Alaska sighting, but that did not necessarily mean that the government actually knew anything at all. "I don't believe the Air Force knows or has any answers," he said.[12]

IN A REVEALING *OMNI* **MAGAZINE INTERVIEW** published a little over a year before his passing, senior editor Pamela Weintraub asked Hynek how his UFO studies informed his religious beliefs, and he readily admitted that, while he felt that the universe was not an accident, it was "an order of magnitude"[13] above his comprehension. Then, perhaps inevitably, he found his own cheeky way to illustrate his point: he invoked a celebrated British astronomer, the late Sir Arthur Eddington, who once observed that, "given the weight of an elephant sliding down a grassy bank and the slope and friction of that grassy bank, a physicist can calculate the exact speed with which the elephant would hit the bottom. But no physicist can tell you why it's funny."[14]

And that was okay. For Hynek, at this moment in his existence, some mysteries were meant to be savored and not necessarily solved.

"He was a completely self-made man who rose to become a well-thought-of astronomer," Paul Hynek recalled of his father. "He found a career that matched his talents, and he wound up— and I don't think there was some grand plan or conspiracy—I think he just happened to be the accidental astronomer.

"And he wound up on this journey to unravel, if not the most, then one of the two or three most important questions to ever face mankind, and he came closer than anyone ever had before. And those kinds of questions don't wrap themselves up neatly in a three-act Hollywood movie. Nor do they typically bookend themselves in the span of one person's natural lifespan. He had to know that, but he brought attention to a subject that he thought merited scientific study, and I think on the whole he felt he did a pretty good job, and that's something to be proud of.

"He bit off something *huge,* and he chewed down a good piece of it."

His dream, Hynek told the *Omni* editor, was to be snowbound on the rocky coast of Maine. "I imagine myself in front of the fireplace, keeping my friends entertained for many nights, not with ghost stories but with one interesting UFO tale after the next. I'd enjoy being given the chance, as long as the food held out."[15]

Hynek knew, of course, that, time and food allowing, he would never run out of those interesting UFO tales, and in one very important sense, he hasn't. One only need watch one of the ubiquitous UFO "documentary" programs on cable TV, or follow along with the triumphant return of Special Agents Mulder and Scully (there's that name again), or insert an "alien-gray" emoji into a text message, or wonder whether one of the seven earthlike planets recently discovered orbiting the red dwarf star TRAPPIST-1 might harbor the life we expected to find on Mars, or count the number of times "close encounter," "Area 51," "little green men," or even "Roswell" pop up in daily conversation, or, as this author has done, spend a few years as a volunteer UFO field investigator for an amateur UFO group, to realize that J. Allen

Hynek has single-handedly brought a consciousness of the UFO phenomenon to the forefront of world culture. We may not understand its meaning or importance yet, but we are all sitting in front of that fireplace in Maine with Dr. Hynek, devouring his stories with an astonishing eagerness and willingness to believe.

There is one fact about UFOs, you see, that Hynek appreciated better than anyone and that causes so many to keep believing. It is a fact that no one can deny and no one can explain: *The phenomenon persists.*

A TIMELINE OF EVENTS IN THE BOOK

1910: Josef Allen Hynek is born to Joseph and Bertha Hynek in North Lawndale, Chicago. When he is five days old, his parents take him out on the roof to bask in the glow of Halley's Comet.

1910: Earth passes through the tail of Halley's Comet, but humanity survives.

1918: Hynek is struck by scarlet fever and reads his first book about astronomy while bedridden.

1924: Hynek's father dies of cardiac failure.

1928: Hynek graduates from high school.

1929: Hynek's mother dies of breast cancer.

1932: Hynek graduates from the University of Chicago with his M.S. in astronomy.

1932: Hynek moves to Yerkes Observatory in Wisconsin to earn his Ph.D. and marries Martha Alexander.

1935: Hynek takes a unique post teaching at both Ohio Wesleyan University and The Ohio State University.

1939: Hynek and his wife divorce.

1942: Hynek marries Mimi Curtis and is drafted to join the war effort, working on the proximity fuze for the first atomic bomb.

1947: Private pilot Kenneth Arnold sees the first "flying saucers" and opens the floodgates of reports.

1948: Air Force Reserve pilot Thomas Mantell crashes and dies while pursuing a massive unidentified flying object, or UFO, over Kentucky.

1948:   The U.S. Air Force recruits Hynek to join its official UFO investigation, Project Sign, to explain away flying saucers as ordinary celestial objects.

1948:   Hynek lies for Project Sign and dreams up a fanciful "atmospheric eddy" to explain UFOs seen by multiple witnesses in Idaho.

1948:   Hynek and Project Sign are stumped when a wingless rocket with banks of lit windows buzzes an Eastern Air Lines passenger plane over Alabama.

1949:   Hynek finishes his Project Sign work, finding 80 percent of flying saucer reports to be misidentifications of natural phenomena; the remaining 20 percent are unexplained.

1949:   Project Sign finds a strong likelihood that those unexplained UFOs are alien spacecraft; Project Grudge takes its place with orders from the Pentagon to squash the extraterrestrial hypothesis.

1950:   The first mass-market UFO book pushes a paranoid conspiracy angle and creates the "saucer crash" myth; it becomes a bestseller.

1951:   *The Thing from Another World* introduces the alien invasion movie genre.

1952:   Project Blue Book takes the place of failed Project Grudge, with a mandate to make the UFO embarrassment go away.

1952:   Hynek is recruited to join Blue Book just as a spectacular mass UFO sighting occurs in the restricted airspace above Washington, D.C.; he is intrigued to find that 20 percent of UFO reports remain unexplained.

1952:   Hynek makes the case for scientific study of UFOs to a conference of scientists, his first public declaration that he has "changed his mind" about the phenomenon.

1953:   The government's Robertson Panel dismisses Hynek's expert testimony and finds the UFO phenomenon not worthy of scientific study.

1954:   Blue Book sends Hynek on his first UFO field investigation of a 1953 mass visual and radar sighting in South and North Dakota, and he finds the sightings credible but unexplainable.

1956:   Hynek takes a leave of absence from Ohio State to develop the world's first global satellite tracking system in anticipation of the United States' historic launch of the first artificial satellite.

1957:   The Soviets beat the United States into orbit with the launch of Sputnik; Hynek becomes the voice of reassurance to Americans jittery about a Soviet attack from space.

1957:   Motorists in and around Levelland, Texas, report seeing a glowing, egg-shaped object that caused their vehicles' engines and headlights to die; police log more than a dozen reports in a single night.

1959:   UFOs are sighted three nights in a row by more than two dozen witnesses in Boianai, Papua New Guinea, and the witnesses interact with the UFO's humanoid occupants.

1960:   Hynek becomes chair of the astronomy department at Northwestern University in Evanston, Illinois.

1960–62: Hynek revolutionizes the science of celestial imaging with the Project Star Gazer high-altitude telescope and the Image Orthicon video telescope.

1961:   Barney and Betty Hill have a terrifying encounter with a UFO and its strange occupants while driving home from their honeymoon; in time, they realize that two hours of their journey are "missing."

1963:   The Hills recall their UFO abduction under hypnosis; Hynek interviews them and finds them credible and convincing.

1964:   Policeman Lonnie Zamora watches a UFO and its occupants fly away, leaving charred bushes at the "landing site"; Hynek declares the New Mexico sighting "unexplained."

1965:   NASA's Mariner 4 Mars mission sends home photos of a dead planet, killing the long-held popular belief that life exists on our closest neighbor.

1965–67: Hynek builds Northwestern's Corralitos Observatory in New Mexico and discovers a record number of supernovae; as a follow-up, he uses his Image Orthicon system to help select landing sites for NASA's Apollo moon missions

1966:   A mass UFO sighting in southern Michigan causes a national sensation; Hynek suggests the witnesses may have seen swamp gas, and the public and media are furious.

1966:   Despite the furor, the swamp gas case makes Hynek an international celebrity; he is vindicated when Congress authorizes a scientific study of UFOs by the University of Colorado.

1969:   The Colorado Project finds that UFOs are not worthy of study and recommends the cancellation of Project Blue Book; the air force happily obliges.

1969:   Now free of the air force's constraints, Hynek speaks out publicly about the reality of the UFO phenomenon and finds support from many other scientists and academics.

1972:   Hynek writes *The UFO Experience,* cementing his position as the world's leading expert on UFOs; the book introduces Hynek's "Close Encounters" classification system for UFO experiences.

1973:   Hynek travels to Papua New Guinea to meet the witnesses from the 1959 "occupants" case, and he finds their story convincing.

1973:   A "flap" of UFO sightings spreads across the nation, culminating in a harrowing abduction in Mississippi and a near midair collision over Ohio between an army helicopter and a UFO.

1973:   In response to the air force's failure to investigate these incidents, Hynek announces the formation of the Center for UFO Studies (CUFOS).

1975:   In his second book, *The Edge of Reality,* Hynek questions the popular extraterrestrial hypothesis and openly wonders whether there is a psychological element to the UFO phenomenon.

1975:   A years-long public argument with Dr. Carl Sagan over the scientific value of UFO research boils over in a passionate live debate between the two.

1977:   Hynek's first book inspires director Steven Spielberg's *Close Encounters of the Third Kind.* Hynek gets a brief cameo in the movie.

1977:   Hynek's third book, *The Hynek UFO Report,* reveals in scathing detail what went wrong with Project Blue Book and the Colorado Project, and maps out the future of UFO research.

1978: CUFOS receives a record number of UFO sighting reports after *Close Encounters of the Third Kind* is released, and the organization creates a computerized database of UFO events. Hynek retires from Northwestern to concentrate on CUFOS.

1980: CUFOS forges ahead with its research projects, even as UFO cases die down and funding declines.

1984: Lured by the promise of lavish funding from a millionaire benefactor, Hynek and his wife move to Scottsdale, Arizona, to set up a new UFO research center affiliated with CUFOS.

1985: Hynek explores the idea that UFOs are part physical and part psychic, and that they come from parallel dimensions.

1985: The promised funding dries up and the new research center fails to materialize. Hynek is diagnosed with prostate cancer, then a brain tumor is discovered.

1986: Hynek passes away as Halley's Comet returns from its seventy-six-year absence.

ACKNOWLEDGMENTS

I freely admit that I was not the first person who recognized that I needed to write this book. When blogging about UFOs led me to the J. Allen Hynek Center for UFO Studies and the opportunity to write a book about Dr. Hynek, my wife, Monica, had a strong hunch that I had just stumbled into my life's work, and in time I realized that she was right (of course I should have known *that* all along). Monica, I can never thank you enough for your wisdom, faith, and support. You've had my back every step of the way and made me feel invincible every time I sat down at the computer to write.

As for the rest of my very lengthy thank-you list, I must begin with the contributions of the late William T. Powers, who for several evenings regaled me with stories of working with Dr. Hynek in the good old days of the 1960s and '70s ("I was the Powers behind the throne," he once said with a chuckle). Bill was thrilled to talk with me about Dr. Hynek and the philosophy of inclusion the two men shared toward the thousands of people who wrote to Hynek about their strange and difficult-to-believe UFO experiences. Bill and Allen believed that if someone had an idea or a story to share, he or she deserved to be listened to, no matter what, and that spirit has guided me throughout the writing of this book. Bill passed away only months after we had the last of our interviews (always done on Skype, because Bill wanted to be face-to-face when we talked), and his brilliance and delightful humor are sorely missed.

Dr. Mark Rodeghier and the volunteers at the J. Allen Hynek Center for UFO Studies in Chicago provided invaluable guidance, advice, and encouragement. There is seemingly not one iota of UFO lore on which Mark is not a font of incredible information, and his colleagues Mary Castner and Frank Reid are amazing historical resources.

Massive thanks must go to former CUFOS stalwart Dr. Michael Swords, who opened his vast private collection of UFO research files to me on more than one occasion and shared many memories of working with Dr. Hynek. Michael's archive of audio recordings of Dr. Hynek's speeches, interviews, and media appearances were priceless, as they introduced me to new elements of Hynek's life and work of which I had not been aware.

Bill Chalker, another old hand at CUFOS, shared many memories from his base in Australia, and he was always ready and willing to answer questions and clarify the ambiguities that are part and parcel of UFO research. As a facilitator of Hynek's fateful 1973 meeting with famous UFO witness Father William Gill, Bill had some unique stories to share about this momentous moment in Hynek's career.

The generosity of UFO historian Dr. David Jacobs cannot be overstated. His friendship with Dr. Hynek spanned many years, and his insights into some of Hynek's most significant choices and motivations were treasured. I am especially grateful to Dr. Jacobs for sharing his memories of learning of Dr. Hynek's ill health and battle with cancer.

I am grateful, too, to Dr. Jacques Vallee for giving me permission to use his published personal journals to help tell Dr. Hynek's story. He felt that his journals, written as they were during Hynek's life, would give me a far more accurate look at his decades-long friendship with Hynek than anything he could share from his "faulty" memories today. I doubt very much that his memories are in any way faulty, but I appreciate his candor and generosity.

I also thank Jennie Zeidman, who started out as Dr. Hynek's

student and became a colleague, confidante, and friend for many decades. Zeidman also granted me permission to use her many writings about her UFO work with Dr. Hynek, for which I am grateful.

There are no words to describe my gratitude to Tina Choate and Brian Myers, who had remained silent for over thirty years about their work with Dr. Hynek in Arizona before taking a huge leap of faith and granting me an interview in 2016. Their attempts to revitalize Hynek's work in the 1980s were ambitious and tumultuous, and even though some grand plans and shining promises collapsed in the end, Hynek remained grateful to Myers and Choate for bringing him to the sunshine of Arizona for his final years.

Paul Hynek's remembrances of his childhood filled in the outline of Hynek's home life with warmth and love. Paul recalled that his dad wasn't always one to bring his work home with him, but when he did, there was no mistaking how strange and wondrous that work was. Paul and his four siblings knew they were growing up in an unconventional household, and they reveled in it! Not only did Paul share many stories about his dad and his family, but he has also continued to share his knowledge and insights throughout the editing process, helping me get the details straight any time I've reached out for help.

Many thanks to Dr. Hynek's colleagues who shared their thoughts with me: Jenny Randles, Jerome Clark, James Oberg, and Dr. Claude Poher all helped me paint a more complete portrait of the man. Thank you, too, to Kathleen Marden and the family of Anthony Ranfone for their permission to use their late relatives' artwork in this book. Finally, a special thank-you to special effects artist Douglas Trumbull, who shared with me his fascinating memories of working on the movie *Close Encounters of the Third Kind,* and who still shares Dr. Hynek's infectious sense of joy and wonder about the mysteries of outer space.

I want to acknowledge the Williams Bay Historical Society in Wisconsin as well as the archivists at Yerkes Observatory, Ohio Wesleyan College, The Ohio State University, the Harvard-

Smithsonian Observatory, and the University of Chicago, all of whom have helped move this project forward in ways both big and small.

Over the past four years the archive room in the basement of Deering Library at Northwestern University in Evanston, Illinois, has become my mothership. Many minor miracles took place in this room: it was here that I discovered Bill Powers, here that I learned of Dr. Hynek's little-known involvement in the space race, and here that I unlocked the full story of Hynek's contentious relationship with Dr. Carl Sagan. This book began and ended at the Northwestern archives, and I will be forever indebted to assistant archivist Janet Olson and archivist Kevin Leonard for all the help and guidance they provided. There are few things more exciting, I've learned, than hearing Janet or Kevin say, "Hey, Mark, I've thought of something you might be interested in."

My agents Megan Close Zavala and Wendy Keller at Keller Media are my heroes. Megan and Wendy realized right away that a "UFO book" didn't just have to be a book about UFOs, and they helped me see the wider potential in the book when I was still thinking small. The fact that Wendy was my very first agent back in the 1990s and made it possible for me to make my first writing sale to *Star Trek: The Next Generation* just makes this whole sequence of events that much more meaningful. And because of Megan's brilliant work, I have had the extreme good fortune to partner with Matthew Daddona, my editor at Dey Street Books, on this project. Matthew admitted to me right away that he was going into this knowing next to nothing about UFOs, and that made him the perfect editor for me. Because of his "convince me" approach to my material, he was always challenging me to make the narrative clearer, more focused, and more accessible. I hope that I have lived up to his high standards.

I am deeply grateful to my children, Nemo, Domino, Calvin, and Capri, and daughter-in-law Stephani for their faith in me. It wasn't that long ago that some of them were good-naturedly scratching their heads over my obsession with UFOs, but they have come around! Y'all don't have to believe in UFOs; I'm just happy that you believe in me. Thank you for coming along with me on this wild ride. My amazing parents, Gene and Jeanne O'Connell, instilled in me early on a love of books and reading and an intense curiosity about the world. From my extravagant college graduation gift of an electric typewriter to last year's trek to Antarctica, you have always encouraged my loftiest dreams and ambitions, and I thank you for that.

Last of all, I'd like to thank the aliens. Without them, this book would be nothing.

For more information on the J. Allen Hynek Center
for UFO Studies, visit www.cufos.org.

NOTES

PROLOGUE

1. S. E. Haydon, "A Windmill Demolishes It," *Dallas Morning News,* April 19, 1897.
2. Ibid.
3. Ibid.
4. "Americana: Close Encounters of a Kind," *Time,* March 12, 1979.
5. William Driskill, letter to Dr. J. Allen Hynek, October 7, 1966.
6. Donald B. Hanlon and Jacques Vallee, "Airships Over Texas," *Flying Saucer Review,* January–February 1967.
7. "A Strange Light on Mars," *Nature,* August 2, 1894.
8. Percival Lowell, *Mars as the Abode of Life* (New York: MacMillan and Company, 1908).
9. William Graves Hoyt, *Lowell and Mars* (Tucson: University of Arizona Press, 1976).
10. Ray Bradbury, interview by students, Chapman College, Orange, CA, 1972.
11. H. G. Wells, *The War of the Worlds* (London: William Heinemann, 1898).
12. Ibid.
13. Brian W. Aldiss, *Billion Year Spree: The True History of Science Fiction* (Garden City, NY: Doubleday, 1973).
14. Ibid.
15. H. G. Wells, "The Things That Live on Mars," *Cosmopolitan,* March 1908.

16. Lilian Whiting, "There Is Life on the Planet Mars," *New York Times,* December 9, 1906.

17. Ibid.

18. "Mars Peopled by One Vast Thinking Vegetable," *Salt Lake Tribune,* October 13, 1912.

19. Ibid.

20. Ibid.

21. "Comet's Poisonous Tail," *New York Times,* February 7, 1910.

22. Gunter Faure and Teresa M. Mensing, *Introduction to Planetary Science: The Geological Perspective* (Dordrecht, Netherlands; London: Springer, 2007).

CHAPTER 1: UNDER THE DOME

1. James Quinlan, "Dr. Hynek: The Rock in 'Scientific Wilderness' of UFOs," *Palm Beach (FL) Post-Times,* November 4, 1973.

2. Ibid.

3. Rudolf Steiner, "The History of Spiritism," lecture, Berlin, May 30, 1904.

4. Donald E. Osterbrock, *Yerkes Observatory, 1892–1950: The Birth, Near Death, and Resurrection of a Scientific Research Institution* (Chicago: University of Chicago Press, 1997).

5. Ian Ridpath, "The Man Who Spoke Out on UFOs," *New Scientist,* May 17, 1973.

6. Ibid.

7. Colin Wilson, *Rudolf Steiner, the Man and His Vision: An Introduction to the Life and Idea of the Founder of Anthroposophy* (Wellingborough, UK: Aquarian Press, 1985).

8. J. Allen Hynek and Jacques Vallee, *The Edge of Reality: A Progress Report on Unidentified Flying Objects* (Chicago: Regnery, 1975).

9. Robert J. Havlik, "A Fair Use of Arcturus: A Syzygy of Scholarians and the Lighting of the Chicago Century of Progress Exposition, 1933–1934," *Journal of Astronomical History and Heritage* 9, no. 1 (2006): 99–108.

10. Philip Kinsley, "Star Sets 1933 Fair Ablaze: Arcturus 'Light Miracle' Thrills Evening Throngs," *Chicago Sunday Tribune,* May 28, 1933.

11. Ibid.

12. Lenox R. Lohry, *Fair Management: The Story of a Century of Progress Exposition—A Guide for Future Fairs* (Chicago: Cuneo Press, 1952).

13. Kinsley, "Star Sets 1933 Fair Ablaze."

CHAPTER 2: UNUSUAL STARS

1. "Nova Herculis, Discovered in December 1934, Varies from First to Thirteenth Magnitudes—Now Fading, About Sixth," *Harvard Crimson*, December 7, 1935.

2. Ridpath, "Man Who Spoke Out on UFOs."

3. Ibid.

4. Quinlan, "Dr. Hynek: The Rock in 'Scientific Wilderness.'"

5. Ridpath, "Man Who Spoke Out on UFOs."

6. "Hynek Lectures at Astronomy Meet," *Ohio Wesleyan Transcript*, March 3, 1936.

7. Ibid.

8. Hadley Cantril, *The Invasion from Mars: A Study in the Psychology of Panic* (Princeton, NJ: Princeton University Press, 1940).

9. A. Brad Schwartz, "The Infamous 'War of the Worlds' Radio Broadcast Was a Magnificent Fluke," Smithsonian.com, May 6, 2015.

10. Jefferson Pooley and Michael J. Socolow, "The Myth of the *War of the Worlds* Panic," *Slate*, October 28, 2013.

11. Cantril, *Invasion from Mars*.

12. "Professors Give Canine Pets Benefits of Higher Education," *Ohio Wesleyan Transcript*, April 26, 1940.

13. Ridpath, "Man Who Spoke Out on UFOs."

14. Pickin' Patter, *Ohio Wesleyan Transcript*, February 6, 1942.

15. Douglas Birch, "'The Secret Weapon of World War II': Hopkins Developed Proximity Fuse," *Baltimore Sun*, January 11, 1993.

16. Ibid.

17. Ibid.

18. *Faculty Member's Annual Report*, Ohio State University Evaluation Program, March 1947.

19. Michael J. Neufeld, *The Rocket and the Reich: Peenemünde and the Coming of the Ballistic Missile Era* (New York: Free Press, 1995).

20. Ibid.

21. Ibid.

22. Ibid.

23. Kenneth Arnold, KWRC radio interview, June 26, 1947.

24. Ibid.

25. Ibid.

26. David Michael Jacobs, *The UFO Controversy in America* (Bloomington: Indiana University Press, 1975).

27. Ibid.
28. National Military Establishment Office of Public Information, "Project 'Saucer,'" memorandum to the press, April 27, 1949.
29. Ibid.

CHAPTER 3: THE CROWDED SKY

1. Statement of Technical Sergeant Quinton A. Blackwell, January 9, 1948, *Project Sign Final Report.*
2. Statement of Private First Class Stanley Oliver, January 9, 1948, *Project Sign Final Report.*
3. U.S. Air Force, *Project Sign Final Report.*
4. Statement of Private First Class Stanley Oliver.
5. Statement of Lieutenant Colonel E. G. Wood, January 9, 1948, *Project Sign Final Report.*
6. Statement of Colonel Guy Hix, January 9, 1948, *Project Sign Final Report.*
7. *Project Sign Final Report.*
8. J. Allen Hynek, "Are Flying Saucers Real?" *Saturday Evening Post,* December 17, 1966.
9. J. Allen Hynek, *The UFO Experience: A Scientific Inquiry* (Chicago: H. Regnery Co., 1972).
10. Hynek, "Are Flying Saucers Real?"
11. Hynek, *UFO Experience.*
12. Ibid.
13. Ibid.
14. Michael Swords, Robert Powell, et al., *UFOs and Government: A Historical Inquiry* (San Antonio, TX; Charlottesville, NC: Anomalist Books, 2012).
15. Jacobs, *UFO Controversy in America.*
16. J. Allen Hynek, *The Hynek UFO Report* (New York: Dell, 1977).
17. National Military Establishment Office of Public Information, "Project 'Saucer'" memo.
18. Ibid.
19. Dennis Stacy, "Close Encounter with Dr. J. Allen Hynek," *MUFON UFO Journal,* February 1985.
20. Hynek and Vallee, *Edge of Reality.*
21. Stacy, "Close Encounter with Dr. J. Allen Hynek."
22. Jacobs, *UFO Controversy in America.*

CHAPTER 4: DEBUNKED

1. Statement of Captain Clarence S. Chiles to S. L. Shannon of Eastern Airlines, August 3, 1948.
2. Albert Riley, "Atlanta Pilots Report Wingless Sky Monster," *Atlanta Constitution,* July 25, 1948.
3. Edward J. Ruppelt, *The Report on Unidentified Flying Objects* (Garden City, NY: Doubleday, 1956).
4. Riley, "Atlanta Pilots Report Wingless Sky Monster."
5. Ibid.
6. Jacobs, *UFO Controversy in America.*
7. Ruppelt, *Report on Unidentified Flying Objects.*
8. Ibid.
9. Jerome Clark, *The UFO Encyclopedia: The Phenomenon from the Beginning,* 2nd ed. (Detroit, MI: Omnigraphics, 1998).
10. *Ohio State University Monthly,* November 1949.
11. Ibid.
12. Ruppelt, *Report on Unidentified Flying Objects.*
13. J. P. Cahn, "The Flying Saucers and the Mysterious Little Men," *True,* September 1952.

CHAPTER 5: SCINTILLATIONS

1. Jacobs, *UFO Controversy in America.*
2. Ibid.
3. "University in Television," *Ohio State University Monthly,* May 1951.
4. "On His Way," *Ohio State University Monthly,* July 1952.
5. Ibid.
6. Jennie Zeidman, "Investigating UFOs—Lessons from a Teacher and Mentor," speech delivered to MUFON of Ohio, November 6, 1999.
7. Ibid.
8. Ruppelt, *Report on Unidentified Flying Objects.*
9. Ibid.
10. Jacobs, *UFO Controversy in America.*
11. Ibid.
12. Ruppelt, *Report on Unidentified Flying Objects.*
13. Ibid.
14. Ibid.
15. Ibid.

16. Ibid.
17. Hynek and Vallee, *Edge of Reality*.
18. Zeidman, "Investigating UFOs."
19. Ruppelt, *Report on Unidentified Flying Objects*.
20. Harry G. Barnes, *Unidentified Targets*, report to the Civil Aeronautics Administration, July 20, 1952.
21. Ibid.
22. Harry G. Barnes, "Radar Man Tells How He Tracked Saucers Over Capital," Newspaper Enterprise Association, August 3, 1952.
23. Ibid.
24. Clark, *UFO Encyclopedia*.
25. Ibid.
26. Robert Emenegger, *UFO's, Past, Present, and Future* (New York: Ballantine Books, 1974).
27. "Many Scientists Sure 'Flying Saucers' Exist," United Press International, July 28, 1952.
28. Ibid.
29. Ibid.
30. Hynek, *UFO Report*.
31. Ruppelt, *Report on Unidentified Flying Objects*.
32. Hynek and Vallee, *Edge of Reality*.
33. Ibid.
34. Hynek, *UFO Report*.
35. Ibid.
36. J. Allen Hynek, Project Blue Book status report, December 1952.
37. Ibid.

CHAPTER 6: PROJECT HENRY

1. Dan Schwartz, "I've Seen a Genuine Film of UFOs That Could NOT Have Been Faked," *National Enquirer*, September 14, 1976.
2. Ruppelt, *Report on Unidentified Flying Objects*.
3. Hynek and Vallee, *Edge of Reality*.
4. Urner Liddel, "Phantasmagoria or Unusual Observations in the Atmosphere," address to the Optical Society of America, October 11, 1952, and published in the *Journal of the Optical Society of America* 43, no. 4 (April 1953).
5. Ibid.
6. Jacobs, *UFO Controversy in America*.
7. Ibid.

8. Ibid.

9. J. Allen Hynek, foreword to *The UFO Handbook: A Guide to Investigating, Evaluating, and Reporting UFO Sightings,* by Allan Hendry (Garden City, NY: Doubleday, 1979).

10. SAC memorandum to FBI headquarters, August 20, 1947.

11. John Brosnan, "Flying Saucer Reported Flashing Down Canyon at 1,000 Miles Per Hour; Two Others Are Seen," *Twin Falls (ID) Times News,* August 15, 1947.

12. Hynek, *UFO Report.*

13. Stacy, "Close Encounter with Dr. J. Allen Hynek."

14. Hynek, *UFO Report.*

15. Timothy Green Beckley, "Exclusive *UFO Report* Interview: Dr. J. Allen Hynek," *Saga UFO Report,* August 1976.

16. Clark, *UFO Encyclopedia.*

17. Emenegger, *UFO's, Past, Present, and Future.*

18. Ruppelt, *Report on Unidentified Flying Objects.*

19. Emenegger, *UFO's, Past, Present, and Future.*

20. Ibid.

21. Ibid.

22. Jennie Zeidman, "I Remember Blue Book," *International UFO Reporter,* March/April 1991.

23. Ibid.

24. Ibid.

25. Ibid.

26. Ibid.

27. Ibid.

28. Ibid.

29. Arthur C. Clarke, book review of *The Conquest of Space, Aeroplane,* January 6, 1950.

30. Ron Miller and Frederick C. Durant III, *Worlds Beyond: The Art of Chesley Bonestell* (Norfolk, VA: Donning, 1983).

31. Willy Ley, "For Your Information: The How of Space Travel," *Galaxy Science Fiction,* October 1955.

32. Martin J. Collins, *After Sputnik: Fifty Years of the Space Age* (New York: Smithsonian Books/HarperCollins, 2007).

CHAPTER 7: HYNEK IN WONDERLAND

1. Staff Sergeant Wesley N. Harry, letter to Dr. J. Allen Hynek, December 9, 1953.

2. Ibid.
3. Bruce D. Callander, "The Ground Observer Corps," *Air Force Magazine,* February 2006.
4. Ruppelt, *Report on Unidentified Flying Objects.*
5. Ibid.
6. Warrant Officer Junior Grade Howell I. Bennett, letter to Dr. J. Allen Hynek, December 17, 1953.
7. Ruppelt, *Report on Unidentified Flying Objects.*
8. Ibid.
9. Ibid.
10. Bennett, letter to Hynek, December 17, 1953.
11. Harry, letter to Hynek, December 9, 1953.
12. Ibid.
13. Ruppelt, *Report on Unidentified Flying Objects.*
14. Ibid.
15. Clark, *UFO Encyclopedia.*
16. Fred Nadis, *The Man from Mars: Ray Palmer's Amazing Pulp Journey* (New York: Jeremy P. Tarcher/Penguin, 2014).
17. Hynek and Vallee, *Edge of Reality.*
18. Ibid.
19. "Astronomer Finds Swedes, Finns to Aid Eclipse Study," Ohio State University Public Relations, October 16, 1953.
20. Ibid.
21. J. Allen Hynek, Scanning the Skies, *Columbus Dispatch,* 1953.
22. Ibid.
23. J. Allen Hynek, public astronomy lecture, Lima, OH, 1953.
24. Cahn, "Flying Saucers and the Mysterious Little Men."
25. H. B. Darrach Jr. and Robert Ginna, "Have We Visitors from Space?" *Life,* April 7, 1952.
26. Ibid.
27. Ibid.
28. Ibid.
29. Patrick Lucanio, *Them or Us: Archetypical Interpretations of Fifties Alien Invasion Films* (Bloomington: Indiana University Press, 1987).
30. Susan Sontag, "The Imagination of Disaster," *Commentary,* October 1, 1965.
31. Ibid.
32. Ibid.

33. Lucanio, *Them or Us.*

34. Major Allen L. Atwell, 740th Aircraft Control and Warning Squadron commander, Ellsworth AFB, letter to Dr. J. Allen Hynek, January 19, 1954.

35. Dr. J. Allen Hynek, letter to Major Allen L. Atwell, 740th Aircraft Control and Warning Squadron commander, Ellsworth AFB, February 2, 1954.

36. "Sun-Gazers," *Ohio State University Monthly,* June 15, 1954.

37. Ibid.

38. J. Allen Hynek, Scanning the Skies, *Columbus Dispatch,* January 3, 1954.

39. Ibid., June 6, 1954.

40. Ibid., July 4, 1954.

41. Ibid., July 15, 1954.

42. Ibid., July 18, 1954.

43. Ibid.

44. Ibid.

CHAPTER 8: FLYING SAUCER CONSPIRACY

1. Arthur Koestler, *The Sleepwalkers: A History of Man's Changing Vision of the Universe,* Part 4: *The Watershed* (London: Hutchinson, 1959).

2. Giorgio de Santillana and Hertha von Dechend, *Hamlet's Mill: An Essay Investigating the Origins of Human Knowledge and Its Transmission Through Myth* (Boston: Godine, 1977).

3. Koestler, *Sleepwalkers.*

4. Colin Wilson, *Starseekers* (Garden City, NY: Doubleday, 1980).

5. Koestler, *Sleepwalkers.*

6. Ibid.

7. U.S. Department of Defense Office of Public Information, "Air Force Releases Study on Unidentified Aerial Objects," press release, October 25, 1955.

8. Dr. J. Allen Hynek, letter to Captain Charles Hardin, December 17, 1955.

9. E. Nelson Hayes, *Trackers of the Skies* (Cambridge, MA: H. A. Doyle Pub. Co., 1968).

10. Ibid.

11. Ibid.

12. Ridpath, "Man Who Spoke Out on UFOs."
13. Lloyd V. Berkner, foreword to *I.G.Y., the Year of the New Moons,* by J. Tuzo Wilson (New York: Knopf, 1961).
14. Ibid.
15. Hayes, *Trackers of the Skies.*
16. Ibid.
17. Walter N. Webb, "Allen Hynek as I Knew Him," *International UFO Reporter,* January/February 1993.
18. Hayes, *Trackers of the Skies.*
19. Ibid.
20. Constance McLaughlin Green and Milton Lomask, *Vanguard: A History* (Washington, D.C.: National Aeronautics and Space Administration, 1970).
21. Webb, "Allen Hynek as I Knew Him."
22. J. Allen Hynek, statement to staff on the first anniversary of Sputnik, 1958.
23. Hayes, *Trackers of the Skies.*
24. Ibid.
25. Ibid.

CHAPTER 9: INTERACTION

1. J. Allen Hynek, speech at the Hypervelocity Impact Conference, Eglin Air Force Base, April 27, 1960.
2. Hynek, *UFO Experience.*
3. Ibid.
4. Clark, *UFO Encyclopedia.*
5. Hynek, *UFO Experience.*
6. *Fort Worth Star-Telegram,* November 4, 1957.
7. Hynek, *UFO Experience.*
8. Ibid.
9. Clark, *UFO Encyclopedia.*
10. Hynek, Hypervelocity Impact Conference speech.
11. Hynek, *UFO Experience.*
12. Ibid.
13. Ibid.
14. Ibid.
15. "Fiery Ovals Sighted in Far Places," *Dallas News,* November 1957.
16. Ibid.
17. Hayes, *Trackers of the Skies.*

CHAPTER 10: OBSCURING INFLUENCE

1. Webb, "Allen Hynek as I Knew Him."
2. Ibid.
3. Clark, *UFO Encyclopedia.*
4. Ibid.
5. Father William Gill, personal notes, June 26, 1959.
6. Allan Hendry and J. Allen Hynek, "Papua/Father Gill Revisited," *International UFO Reporter,* November 1977.
7. Gill, personal notes.
8. Norman E. G. Cruttwell, *Flying Saucers over Papua: A Report on Papuan Unidentified Flying Objects,* 1960.
9. Ibid.
10. Hynek, *UFO Experience.*
11. Ibid.
12. Ibid.
13. Ridpath, "Man Who Spoke Out on UFOs."
14. Dr. J. Allen Hynek, memorandum to Dr. Fred Whipple, December 14, 1959.
15. Dr. Fred Whipple, letter to Dr. J. Allen Hynek, September 9, 1960.
16. Ridpath, "Man Who Spoke Out on UFOs."
17. J. Allen Hynek, "Occultation of the Bright Star Regulus by Venus," *Science,* September 18, 1959.
18. "Dr. J. Allen Hynek," undated Northwestern University staff biography.
19. Ibid.
20. Ibid.
21. J. Allen Hynek, *Scientific Report on Project Star Gazer: Final Technical Report,* July 1966.
22. Ibid.
23. J. Allen Hynek, "Man's Satellites: Doorway to Space," Thirtieth Steinmetz Memorial Lecture, broadcast on WRGB-TV, Schenectady, NY, May 17, 1958.
24. Ibid.
25. Ridpath, "Man Who Spoke Out on UFOs."
26. J. R. Dunlap, J. A. Hynek, and W. T. Powers, "Improvements in the Application of the Image Orthicon to Astronomy," presented at the Fifth Annual Symposium on Photo-Electronic Image Devices, Imperial College, and published in *Advances in Electronics and Electron Physics* 33, pt. B (1972): 789–94.

27. "Hynek," undated Northwestern University staff biography.
28. Ridpath, "Man Who Spoke Out on UFOs."
29. "N.U. Dedicates Observatory at Texas–New Mexico Border," *Evanston Review,* April 13, 1961.
30. Stephen King, *Danse Macabre* (New York: Everest House, 1981).
31. Joseph Stefano, interview by Lee Weinstein, 1975, published in *New Variant,* no. 2 (1983).
32. Kathleen Marden and Stanton T. Friedman, *Captured! The Betty and Barney Hill UFO Experience: The True Story of the World's First Documented Alien Abduction* (Franklin Lakes, NJ: New Page Books, 2007).
33. John G. Fuller, *The Interrupted Journey: Two Lost Hours "Aboard a Flying Saucer"* (New York: Dial Press, 1967).
34. Ibid.
35. Ibid.
36. Ibid.
37. Ibid.

CHAPTER 11: BURNED BRUSH

1. "Did Creature from Space Burn Grass?" United Press International, April 26, 1964.
2. Lonnie Zamora, interview by Captain Richard T. Holder, U.S. Army, April 24, 1964.
3. Ibid.
4. Ibid.
5. Dr. J. Allen Hynek, personal notes on remarks, May 5, 1964.
6. Zamora, interview by Holder.
7. Hynek, notes on remarks.
8. Captain Hector Quintanilla, memo to Colonel Eric T. de Jonckheere, April 27, 1964.
9. Zamora, interview by Holder.
10. "Evidence of UFO Landing Observed," *El Defensor Chieftain* (Socorro, NM), April 28, 1964.
11. Ibid.
12. Hynek, notes on remarks.
13. "Evidence of UFO Landing Observed."
14. Hynek, notes on remarks.
15. "Air Force Consultant Checks UFO Site Here," *El Defensor Chieftain* (Socorro, NM), April 30, 1964.
16. Hynek, notes on remarks.

17. Ibid.

18. Ibid.

19. Ibid.

20. Dr. J. Allen Hynek, letter to Dr. Donald Menzel, September 29, 1964.

21. Coral Lorenzen, "UFO Lands in New Mexico," *Fate,* August 1964.

22. Ibid.

23. "Flying 'Thing' Radar-Proof?" Associated Press, April 30, 1964.

24. Dr. J. Allen Hynek, interview by Walter Shrode, broadcast on KSRC, April 29, 1964.

25. "Flying 'Thing' Radar-Proof?"

26. "Areas Trampled—Curious Citizenry Foils Investigator," Associated Press, April 29, 1964.

27. Walter Webb, NICAP report, November 4, 1964.

28. Hynek, letter to Menzel, September 29, 1964.

29. Hynek, *UFO Report.*

30. Hynek, notes on remarks.

31. Jacques Vallee, *Forbidden Science: Journals 1957–1969* (New York: Marlowe & Company, 1996).

32. Ibid.

33. J. Allen Hynek, *Scientific Report on Project Star Gazer: Final Technical Report,* July 1966.

34. Ibid.

35. Ibid.

36. Ibid.

37. Ibid.

38. "Astronomer Wants Lens on Moon," *Kansas City Star,* March 3, 1965.

39. Frank Hughes, "N.U. Helps United States Aim for Moon," *Chicago Tribune,* February 19, 1964.

40. Elizabeth Howell, "Mariner 4: First Spacecraft to Mars," Space.com, December 5, 2012.

CHAPTER 12: WILL-O'-THE-WISP, PART ONE

1. Jim Fiebig, "Washtenaw Farmer Tells of UFO," *Port Huron Times Herald,* March 22, 1966.

2. Mort Young, "They're Still Seeing Things in the Sky," *New York Journal-American,* March 27, 1966.

3. Fiebig, "Washtenaw Farmer Tells of UFO."

4. Norman Glubock, "Saucertown, U.S.A.—Weird Week in Dexter," *Chicago News,* March 26, 1966.

5. Vivian M. Baulch, "The Great Michigan UFO Chase," *Detroit News,* February 14, 1995.

6. Young, "They're Still Seeing Things in the Sky."

7. Gidget Kohn, "Hillsdale Coeds Spot Flying Saucer; Eye-Witness Account," *Hillsdale Collegian,* March 24, 1966.

8. Ibid.

9. Young, "They're Still Seeing Things in the Sky."

10. Howard Fields, untitled article, United Press International, March 22, 1966.

11. Kohn, "Hillsdale Coeds Spot Flying Saucer."

12. Kelly Rae Hearn, statement issued to the press, March 27, 1966.

13. Kohn, "Hillsdale Coeds Spot Flying Saucer."

14. Hearn, statement issued to the press.

15. Kohn, "Hillsdale Coeds Spot Flying Saucer."

16. Fields, untitled article.

17. Kohn, "Hillsdale Coeds Spot Flying Saucer."

18. Hynek, *UFO Report.*

19. Swords, Powell, et al., *UFOs and Government.*

20. Vallee, *Forbidden Science: Journals 1957–1969.*

21. Ibid.

22. Hynek, "Are Flying Saucers Real?"

23. Ibid.

24. Hynek and Vallee, *Edge of Reality.*

25. Vallee, *Forbidden Science: Journals 1957–1969.*

26. Hynek, "Are Flying Saucers Real?"

27. Hynek and Vallee, *Edge of Reality.*

28. United Press International, March 23, 1966.

29. Ibid.

30. "Set Up Night Watch for Michigan Lights," *Chicago Tribune,* March 23, 1966.

31. "Air Force to Explain 'Saucers,'" *Detroit Free Press,* March 25, 1966.

32. Swords, Powell, et al., *UFOs and Government.*

33. Hynek, "Are Flying Saucers Real?"

34. Hector Quintanilla, "UFOs: An Air Force Dilemma" (unpublished manuscript, 1975), available at the National Institute for Discovery Science.

35. Ibid.

CHAPTER 13: WILL-O'-THE-WISP, PART TWO

1. Hynek, "Are Flying Saucers Real?"
2. Paul O'Neil, "A Well-Witnessed 'Invasion'—by Something," *Life*, April 1, 1966.
3. Vallee, *Forbidden Science: Journals 1957–1969*.
4. Hynek, "Are Flying Saucers Real?"
5. "Air Force to Explain 'Saucers.'"
6. Ibid.
7. Vallee, *Forbidden Science: Journals 1957–1969*.
8. Jacobs, *UFO Controversy in America*.
9. Quintanilla, "UFOs: An Air Force Dilemma."
10. *Unidentified Flying Objects: Hearing,* by Committee on Armed Forces of the House of Representatives, 89th Congress, 2nd Session (April 5, 1966), p. 6073.
11. Hynek and Vallee, *Edge of Reality*.
12. David R. Saunders and R. Roger Harkins, *UFOs? Yes! Where the Condon Committee Went Wrong* (Toronto: Signet Books, 1968).
13. Vallee, *Forbidden Science: Journals 1957–1969*.
14. Quintanilla, "UFOs: An Air Force Dilemma."
15. Swords, Powell, et al., *UFOs and Government*.
16. Hynek and Vallee, *Edge of Reality*.
17. Vallee, *Forbidden Science: Journals 1957–1969*.
18. Dr. Thornton Page, interview by Antonio Huneeus, June 22, 1983.
19. Vallee, *Forbidden Science: Journals 1957–1969*.
20. Ann Druffel, *Firestorm: Dr. James E. McDonald's Fight for UFO Science* (Columbus, NC: Wild Flower Press, 2003).
21. Hynek and Vallee, *Edge of Reality*.
22. Ibid.
23. Ibid.
24. J. Allen Hynek, letter to the editor, *Science*, October 21, 1966.
25. Ibid.
26. Ibid.
27. Hynek, "Are Flying Saucers Real?"

CHAPTER 14: MR. UFO

1. Fuller, *Interrupted Journey*.
2. Ibid.
3. Ibid.

4. Ibid.
5. Ibid.
6. Ibid.
7. Ibid.
8. Ibid.
9. Ibid.
10. Hynek, *UFO Experience.*
11. Dr. J. Allen Hynek, speech to Midwest Hypnosis Convention, 1979.
12. Ibid.
13. Hynek, *UFO Experience.*
14. Saunders and Harkins, *UFOs? Yes!*
15. Ibid.
16. Vallee, *Forbidden Science: Journals 1957–1969.*
17. Ibid.
18. Ibid.

CHAPTER 15: SIGNAL IN THE NOISE

1. Saunders and Harkins, *UFOs? Yes!*
2. Ibid.
3. Ibid.
4. J. Allen Hynek, *Follow-up Report on Socorro Case,* March 1965.
5. "Moment of Truth for J. Allen Hynek, We Hope!" *Journal of Borderland Science Research* 22, no. 7 (October 1966).
6. Saunders and Harkins, *UFOs? Yes!*
7. Ibid.
8. Ibid.
9. Jacobs, *UFO Controversy in America.*
10. Dr. J. Allen Hynek, "Project Blue Book: Its History and Methods," presented to the Colorado Project, November 11, 1966.
11. Saunders and Harkin, *UFOs? Yes!*
12. Hynek and Vallee, *Edge of Reality.*
13. Major Hector Quintanilla, letter to Dr. J. Allen Hynek, October 11, 1968.
14. Dr. J. Allen Hynek, letter to Major Hector Quintanilla, October 21, 1968.
15. Dr. J. Allen Hynek, Cacciopo and Sweeney discussion notes, February 15, 1968.
16. Dr. J. Allen Hynek, letter to John Fuller, December 6, 1967.
17. J. Allen Hynek, "The UFO Gap," *Playboy,* December 1967.

18. Ibid.
19. Ibid.
20. David Butler, assistant editor, *Playboy,* letter to Dr. J. Allen Hynek, January 5, 1968.
21. Laurence A. Marschall, *The Supernova Story* (New York: Springer Science+Business Media, 1988).
22. Richard Lewis, "The Flying Saucer Man," *Chicago Sun-Times,* December 22, 1966.
23. Ibid.
24. Ibid.
25. Ibid.
26. Ibid.

CHAPTER 16: INVISIBLE AT LAST

1. Jacques Vallee, *The Invisible College: What a Group of Scientists Has Discovered About UFO Influences on the Human Race* (New York: Dutton, 1975).
2. Ibid.
3. Andrew Tomas, *We Are Not the First: Riddles of Ancient Science* (London: Souvenir Press, 1971).
4. Hynek and Vallee, *Edge of Reality.*
5. Ibid.
6. Saunders and Harkins, *UFOs? Yes!*
7. Ibid.
8. Hynek, *UFO Report.*
9. J. Allen Hynek, "The Condon Report and UFOs," *Bulletin of the Atomic Scientist,* April 1969.
10. Hynek, *UFO Report.*
11. Ibid.
12. Vallee, *Forbidden Science: Journals 1957–1969.*
13. James Renner, "Strangers in the Night," *Cleveland Scene,* March 31, 2004.
14. Ibid.
15. Ibid.
16. Swords, Powell, et al., *UFOs and Government.*
17. Hynek, *UFO Experience.*
18. James M. Peek, Sandia Corporation, letter to Dr. J. Allen Hynek, January 20, 1967.
19. Dr. Carl Sagan, letter to Dr. J. Allen Hynek, September 29, 1969.

20. Dr. J. Allen Hynek, letter to Dr. Carl Sagan, October 27, 1969.

21. J. Allen Hynek, "Twenty-One Years of UFO Reports," UFO symposium, American Association for the Advancement of Science, 134th Meeting, December 27, 1969.

22. Carl Sagan and Thornton Page, *UFO's—A Scientific Debate* (Ithaca, NY: Cornell University Press, 1972).

23. Ibid.

24. Ibid.

25. Karen Masters, "How Many Known Galaxies Are There?" Ask an Astronomer, November 19, 2000 (last updated July 18, 2015), http://curious.astro.cornell.edu/95-the-universe/galaxies/general-questions/516-how-many-known-galaxies-are-there-intermediate.

26. Elizabeth Howell, "How Many Stars Are in the Universe?" Space.com, May 31, 2014.

27. Ibid.

28. National Aeronautics and Space Administration, "NASA's Kepler Mission Announces Largest Collection of Planets Ever Discovered," press release, May 10, 2016.

29. Ibid.

30. Hynek, "Twenty-One Years of UFO Reports."

31. Ibid.

32. Ibid.

CHAPTER 17: THE UFO EXPERIENCE

1. J. Allen Hynek, "It's a Bird, It's a Plane, It's . . ." *Christian Science Monitor,* April 22, 1970.

2. Ibid.

3. Ibid.

4. Vallee, *Forbidden Science: Journals 1957–1969.*

5. Calendar listing, *Daily Northwestern,* February 22, 1977.

6. Hynek, *UFO Experience.*

7. Ibid.

8. Ibid.

9. Ibid.

10. Ibid.

11. Ibid.

12. Ibid.

13. Ibid.

14. Ibid.

15. Ibid.

16. Quintanilla, "UFOs: An Air Force Dilemma."

17. Hynek, *UFO Experience.*

18. Quintanilla, "UFOs: An Air Force Dilemma."

19. Hynek, *UFO Experience.*

20. Ibid.

21. Ibid.

22. Quintanilla, "UFOs: An Air Force Dilemma."

23. David L. Miller, *Introduction to Collective Behavior and Collective Action,* 3rd ed. (Prospect Heights, IL: Waveland Press, 2014).

24. Bruce C. Murray, "Reopening the Question," review of J. Allen Hynek's *The UFO Experience, Science,* August 25, 1972.

25. Dr. J. Allen Hynek, "Update on the UFO Situation," speech, 1973.

26. Caren Marcus, "Hynek Traces UFO Facts," *Daily Northwestern,* April 30, 1973.

27. Hynek, *UFO Report.*

28. Ibid.

29. Hynek, "Update on the UFO Situation."

30. Hynek and Vallee, *Edge of Reality.*

31. Hynek, *UFO Report.*

32. Stephanie Schorow, "Hynek calls UFOs 'Authentic,'" *Daily Northwestern,* September 21, 1973.

33. Ibid.

34. Ibid.

35. Jacobs, *UFO Controversy in America.*

36. Ibid.

CHAPTER 18: THE SPUR

1. Calvin Parker Jr. video statement, n.d.

2. Charles Hickson, interview by Jackson County sheriff Fred Diamond and Captain Glen Ryder, October 11, 1973.

3. Charles Hickson and William Mendez, *UFO Contact at Pascagoula* (Tucson, AZ: self-published, 1983).

4. Ibid.

5. Ibid.

6. Charles Hickson video interview, n.d.

7. Parker video statement.

8. Ibid.
9. Transcript of secret recording of Charles Hickson and Calvin Parker Jr. by Sheriff Fred Diamond, October 12, 1973.
10. Hickson, video interview.
11. Ibid.
12. Ibid.
13. Hickson and Mendez, *UFO Contact at Pascagoula.*
14. Ralph Blum, "We Were Captured by Aliens," *National Enquirer,* n.d.
15. J. Allen Hynek, WWDC radio interview, 1974.
16. Hynek, *UFO Experience.*
17. Beckley, "Exclusive *UFO Report* Interview: Dr. J. Allen Hynek."
18. Ibid.
19. Ralph Blum and Judy Blum, *Beyond Earth: Man's Contact with UFOs* (New York: Bantam Books, 1974).
20. Ibid.
21. Ibid.
22. Murphy Givens, "NASA to Probe," *Mississippi Press,* October 15, 1973.
23. Maren Rudolph, "Scientific Opinions Vary on 'Sightings,'" *New Orleans Times-Picayune,* October 21, 1973.
24. Murphy Givens, "Incident Is Retraced Week After Happening," *Mississippi Press,* October 18, 1973.
25. "UFOs: Invitation to Discovery?" *Argonne News,* November 1973.
26. Lawrence Coyne, Arrigo Jezzi, John Healey, and Robert Yanacsek, *Near Midair Collision with UFO Report,* Eighty-Third U.S. Army Reserve Command, Columbus, OH, November 23, 1973.
27. Ibid.
28. Ibid.
29. "Sighting: The Strange Case of Major Coyne," *Army Reserve Magazine,* September–October 1974.
30. Ibid.
31. Ibid.
32. Clark, *UFO Encyclopedia.*

CHAPTER 19: PURPLE PEACH TREES

1. Jennie Zeidman, "J. Allen Hynek: A 'Rocket Man,'" MUFON of Ohio, 1999, http://www.mufonohio.com/mufono/hynek.html.
2. Blum and Blum, *Beyond Earth.*
3. Ibid.
4. Ibid.

5. Ibid.

6. Ibid.

7. Ibid.

8. Jacques Vallee, *Forbidden Science: Journals 1970–1979* (San Francisco, CA: Documatica Research, 2008).

9. Blum and Blum, *Beyond Earth.*

10. Ibid.

11. Hynek, WWDC radio interview.

12. Dennis V. Waite, "Hello, Out There," *Chicago Sun-Times,* December 30, 1973.

13. Ibid.

14. Ibid.

15. Ibid.

16. Jennie Zeidman, *Report on Coyne Helicopter Incident,* Center for UFO Studies, 1979.

17. George Gallup, "They Saw UFOs, 15 Million Say," press release, American Institute of Public Opinion, November 1973.

18. David M. Jacobs, letter to Dr. J. Allen and Mimi Hynek, December 24, 1974.

19. Ibid.

20. David Galanti, "Hynek on the Track of UFO Research," *Daily Northwestern,* February 2, 1974.

21. Dr. J. Allen Hynek, letter to Vice President of Development John E. Fields, Northwestern University, November 18, 1974.

22. Ibid.

23. Ibid.

24. Northwestern University provost Raymond Mack, letter to Dr. J. Allen Hynek, December 9, 1974.

25. Howard S. Goller, "Profs Ditch Classes for TV Prime Time," *Daily Northwestern,* May 1, 1974.

26. Ibid.

27. *Summer Northwestern,* August 19, 1977.

28. "'Martian' Streaker Is Green," *Chicago Sun-Times,* May 29, 1970.

29. Dr. J. Allen Hynek, letter to Dr. Carl Sagan, October 28, 1975.

CHAPTER 20: HYNEK VS. SAGAN

1. "Science in Collision: The Hynek-Sagan UFO Forum," *UFO Magazine,* February 1980.

2. Ibid.

3. Ibid.
4. Ibid.
5. Ibid.

CHAPTER 21: CLOSE ENCOUNTERS

1. Claude Poher, *Some Studies and Reflection About the U.F.O. Phenomenon,* July 1974.
2. Ibid.
3. Ibid.
4. Clark, *UFO Encyclopedia.*
5. Ibid.
6. Hynek and Vallee, *Edge of Reality.*
7. Ibid.
8. Ibid.
9. Ibid.
10. Ibid.
11. Ibid.
12. Dr. J. Allen Hynek, letter to Steven Spielberg, January 8, 1976.
13. Ibid.
14. Ibid.
15. Miriam Conrad, "Hynek Encounters Stars in Sky and Hollywood," *Daily Northwestern,* January 6, 1978.
16. David Gerrold, "*Close Encounters* and *Star Wars,*" *Science Fantasy Film Classics,* Spring 1978.
17. Jeff Bloch, "Eye-in-the-Sky Hynek—Intellectual Superstar?" *Daily Northwestern,* May 3, 1978.
18. Ibid.
19. Ibid.
20. Ibid.
21. Ibid.
22. Ibid.
23. Ibid.
24. Hynek, *UFO Report.*
25. Ibid.
26. Dr. J. Allen Hynek, KGRO radio interview, Boston, MA, 1981.

CHAPTER 22: ARIZONA

1. Jeff Lyon, "Hynek Ends as He Began: Getting Students to Think," *Chicago Tribune,* June 6, 1978.

2. Jeff Bloch, "Profs Bid Farewell," *Daily Northwestern,* May 31, 1978.

3. J. Allen Hynek, "UFO Newsfront," *International UFO Reporter,* July 31, 1977.

4. "First Summary of the Work of the French Government's 'GEPAN' UFO Organization," *International UFO Reporter,* October/November 1978.

5. Ibid.

6. Mark Gelatt, "UFO Center Is Looking Outward," *Daily Northwestern,* October 21, 1980.

7. Val Johnson, Minnesota Public Radio interview, October 28, 2015.

8. Allan Hendry, "Val Johnson: A Personal Profile," *International UFO Reporter,* September/October 1979.

9. Ibid.

10. John Enger, "Whatever Happened to the Marshall County Cop Who Hit a UFO?" MPR News, August 27, 2015.

11. Joanna Robinson, "The Cast of *Fargo* Reacts to This Season's Most Bonkers Plot," *Vanity Fair,* October 20, 2015.

12. Michael Sperling, "A Close Encounter with J. Allen Hynek," *Daily Northwestern,* February 17, 1982.

13. Ibid.

14. Ibid.

15. J. Allen Hynek, editorial, *International UFO Reporter,* May/June 1982.

16. Ibid.

17. Jerome Clark, "UFOs Still Here—Sort Of," *International UFO Reporter,* March/April 1986.

18. Ibid.

19. David Dreier, "A Close Encounter with Professor Hynek," *North Shore,* December 1980.

20. J. Allen Hynek, *Psychic Phenomena* television interview, 1979.

21. Ibid.

22. J. Allen Hynek, "Converting the Disbelievers," interview by Pamela Weintraub, *Omni,* February 1985.

23. Howard Witt, "UFO Expert Moving to Arizona," *Chicago Tribune,* August 21, 1984.

24. Ibid.

25. Hynek, "Converting the Disbelievers."

26. Ibid.

27. Ibid.

28. Ibid.

29. Ibid.

30. J. Allen Hynek, "UFOlogy as a Profession: A Manifesto," *CUFOS Associate Newsletter,* September 1981.

31. Dr. Willy Smith, statement of position, April 1, 1992.

32. J. Allen Hynek, "Why UFOs?" *International UFO Reporter,* January/February 1985.

33. Jacques Vallee, *Forbidden Science: Journals 1980–1989* (San Francisco, CA: Documatica Research, 2016).

34. "Dr. Hynek Resigns," *International UFO Reporter,* November/December 1985.

35. Vallee, *Forbidden Science: Journals 1980–1989.*

36. Ibid.

37. Ibid.

38. Ibid.

39. Hynek, "Why UFOs?"

40. "Dr. Hynek Resigns."

41. Editorial, *International UFO Reporter,* September/October 1986.

CHAPTER 23: THE SUPERSENSIBLE REALM

1. Dr. J. Allen Hynek and Dr. Willy Smith, letter to Marge Christensen, August 25, 1985.

2. Hynek, "Converting the Disbelievers."

3. Vallee, *Forbidden Science: Journals 1980–1989.*

4. Ibid.

5. Rudolf Steiner, *An Outline of Occult Science* (New York: Anthroposophic Press, 1920).

6. Jennie Zeidman, "The Comet and the Circle," *International UFO Reporter,* May/June 1986.

7. Ibid.

8. Ibid.

9. Ibid.

10. Paul Fanning, *City Elements* interview, WCFL Radio/University of Illinois, 1980.

11. "Conversation with Dr. J. Allen Hynek," *Oui,* April 1977.

12. Ibid

13. Hynek, "Converting the Disbelievers."

14. Ibid.

15. Ibid.

INDEX

Bold page numbers contain illustrations.

ABOUT THE AUTHOR

Mark O'Connell is a screenwriter, teacher, and blogger. He wrote episodes for *Star Trek: The Next Generation* and *Star Trek: Deep Space Nine,* and has developed feature film projects with major studios, including Walt Disney and DreamWorks Animation. He is also the founder of the UFO blog *High Strangeness*. He lives in Wisconsin with his wife, Monica, and teaches screenwriting at DePaul University in Chicago.